FIELD MANUAL
FM 5-20

HEADQUARTERS,
DEPARTMENT OF THE ARMY
WASHINGTON, D.C., *May 20 1968*

CAMOUFLAGE

			Paragraph	Page
CHAPTER	1.	INTRODUCTION	1–4	3
	2.	DETECTION	5–8	4
	3.	CONCEALMENT	9,10	14
	4.	CAMOUFLAGE FOR THE INDIVIDUAL	11–17	26
	5.	FIELD FORTIFICATIONS	18–22	31
	6.	VEHICLES AND ARTILLERY		
Section	I.	Vehicles	23,24	35
	II.	Artillery	25,26	44
CHAPTER	7.	AIRCRAFT	27–31	49
	8.	BIVOUACS, COMMAND POSTS AND SUPPLY POINTS	32–38	57
	9.	LARGE SCALE AND RELATIVELY PERMANENT INSTALLATIONS	39–55	66
	10.	DECOY INSTALLATIONS	56–74	86
APPENDIX		REFERENCES		102
INDEX				103

* This manual supersedes FM 5-20, 21 January 1959; FM 5-21, 23 December 1958; and together with TM 5-200, 25 September 1959 supersedes FM 5-23, 3 October 1956, and all changes.

CHAPTER 1

INTRODUCTION

1. Purpose and Scope

a. This manual provides a comprehensive reference and guide in all aspects of camouflage. It describes in detail the principles involved in concealing or disguising troops, vehicles, weapons, and field installations. The last chapter discusses the planning for and the operation of decoys and decoy installations. For technical information regarding the natural and artificial materials and equipment available for both concealment and visual deception, the reader is referred to TM 5–200.

b. Information and data presented in this manual are applicable to both nuclear and nonnuclear warfare. It must be remembered that in the event of a nuclear burst near a camouflaged position, the thermal radiation may ignite the camouflage if it is of a flammable substance. In addition, radiological decontamination operations may be hindered by the presence of elaborate camouflage construction. These are calculated risks and must be taken into consideration when planning camouflage measures.

2. Comment

Users of this manual are encouraged to submit recommended changes or comments to improve the manual. Comments should be keyed to the specific page, paragraph, and line of the text in which the change is recommended. Reasons should be provided for each comment to insure understanding and complete evaluation. Comments should be forwarded directly to the Commandant, U. S. Army Engineer School, Fort Belvoir, Va., 22060.

3. Responsibilities for Camouflage and Deception

a. Individual. The individual soldier is responsible for his own concealment. His responsibility here is just as great as his responsibility for his rifle, and he must know as much about camouflage as he does about his weapon. Just as training in marksmanship teaches the soldier to hit a target accurately, so does knowledge of camouflage teach him how to escape becoming a target himself.

b. Command. Overall unit camouflage is the responsibility of the commander. He must insure the complete understanding by every man in his command of the importance, principles, and techniques of camouflage. In addition to his responsibilities in the training and supervision of the individual soldier's concealment efforts, he must plan and execute camouflage measures for the operational, administrative, and logistical areas of his command. Finally, to insure the effectiveness of all camouflage measures, he is responsible for the strict enforcement of camouflage discipline.

4. Importance of Camouflage

Camouflage is one of the basic weapons of war. Correctly used, it can spell the difference between a successful campaign and defeat; to the individual it can mean the difference between life and death. Regardless of the type of warfare—all out nuclear or internal defense operations—camouflage remains important. Small semi-independent units must furnish their own security, reconnaissance, and surveillance. They must be able to exist for long periods of time with a minimum of control and support from higher headquarters. As a result, their success will depend to a large extent upon their ability to remain concealed from the enemy. This in turn will depend upon the knowledge and proper execution of the principles of camouflage.

CHAPTER 2

DETECTION

5. Introduction

Before it is possible to employ effective camouflage it is necessary to know something about observation. There are two broad categories of observation—direct and indirect.

6. Direct Observation

Direct observation refers to that process whereby the observer sees the subject physically, that is, with his eyes—aided or unaided. Examples of this type of observation include an observer sitting on a hilltop with binoculars or an aerial observer viewing the landscape from an aircraft. Direct observation has many advantages: It offers immediate information on which action may be based; the picture is seen in the true third dimension and is easily evaluated by the brain; the eye is normally an accurate and sensitive receiver; and it allows observation of movement. It has four major disadvantages: There is no permanent record for future direct comparison; weather and time of day may limit it; the observer's experience and mission may limit the information obtained; and human error may result in incomplete and incorrect information.

7. Indirect Observation

In indirect observation the observer sees a picture or an image of the subject, and not the physical subject itself. Photography, radar, infrared, and television are used in indirect observation. The advantages of indirect obervation are: Successive photos of the same area may detect changes which have taken place; it results in a permanent record; it increases the spectrum for observation to the infrared and other spectrums invisible to the human eye; it can be distributed to all echelons for their particular needs; the photographs may be studied at length; it reduces distance of observation by enlargement; and radar will usually detect a metallic object behind a nonmetallic screen, such as a conventional burlap-garnished camouflage net. Its disadvantages are: The time required to develop photographs and disseminate the information; weather may interfere with aerial photograph; the skill of the photointerpreter may limit information obtained; and there is a lack of the third dimension, except on special stereophotography.

a. Aerial Photography. In modern warfare aerial photography has assumed a place of extreme importance, and in regard to camouflage detection and inspection, photography has reached a stage where it is indispensable. Military photographs are divided into three categories: vertical, high oblique, and low oblique.

(1) *Vertical.* The vertical photograph is one taken directly above the subject. It shows practically no detail in the third dimension other than shadow and can be compared with a plan view of buildings on a blueprint. When taking a vertical photograph, the line of sight on the camera is perpendicular to the line of flight of the aircraft (fig. 1). In vertical photo interpretation, the process of stereovision is used extensively. By taking two photographs of the same subject and stereoscopic glasses, the third dimension, depth, becomes apparent. This is of great advantage in examining enemy camouflage or in inspecting our own camouflage.

(2) *High oblique.* High oblique photographs are those taken at an angle raised from the vertical so that the apparent horizon shows on the photo. It also shows a partial third

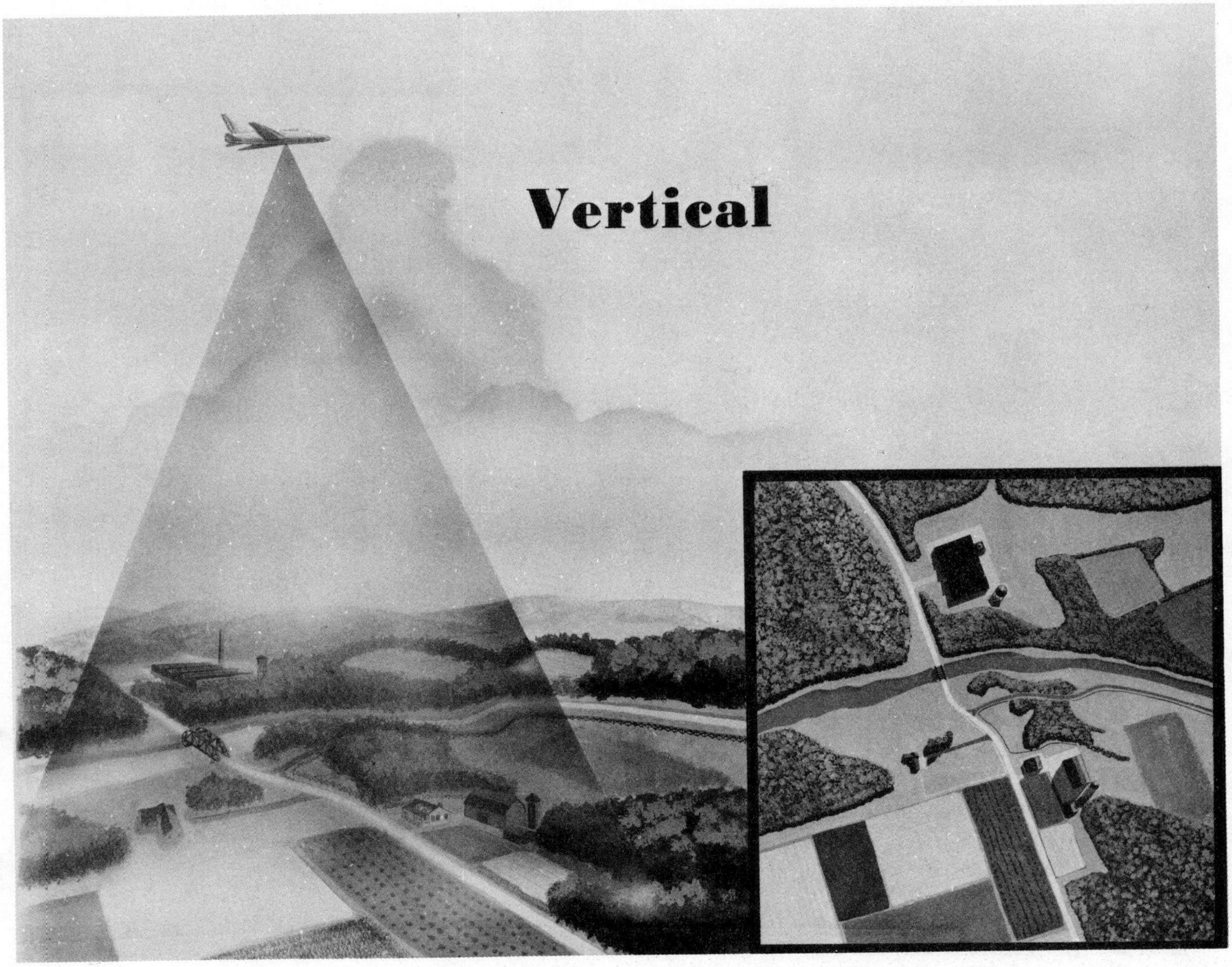

Figure 1. Vertical photograph.

dimension by giving the side and top view of the subject (fig. 2).

(3) *Low oblique.* The low oblique is similar to the high oblique except that it does not show the horizon. This is accomplished by taking the photograph at an angle less than 30° from the perpendicular to the line of flight (fig. 3).

(4) *Films.* There are four types of film used in aerial photography:

(*a*) *Black and white.* Black and white film records images in tone gradation between white and black. While not reproducing color, it does provide a permanent record of tonal differences. Select filters are often used to improve the photograph or to record only the light that is known to give the greatest tonal differences between natural backgrounds and the object being sought (fig. 4).

(*b*) *Color film.* Color film will detect camouflage which does not match the colors in the background. However, this film has many operational difficulties and gives best results only under ideal conditions. Colors tend to blend together at high altitudes, shadow density is not as accurate on this film as on the black and white film, and atmospheric conditions must be ideal in order to obtain a clear photograph. For these reasons, this film is not widely used (fig. 5).

(*c*) *Infrared film.* Infrared waves or rays refer to a portion of the electromagnetic spectrum which is invisible to the human eye. Most things in nature, such as living, green vegetation, reflect these infrared waves readily and in large quantities. Most artificial ma-

Figure 2. High oblique photograph.

terials normally do not reflect these infrared waves to the same extent. Thus, infrared film which is nothing more than black and white film that is sensitive to infrared waves, can result in a picture showing contrasts between natural materials and artificial materials. The natural materials will show up as a light tone of gray while the artificial materials will show up as a dark tone of gray. Infrared film has another important use. It can be used to take photographs at night if there is a source of infrared radiation. To counter the detecting ability of this film, camouflage paints and dyes have been developed that have a high infrared reflectance, similar to foliage. All camouflage materials are now issued with this type of coloring so that infrared film can no longer detect differences between natural and artificial camouflage that has been treated with such paint and dye (fig. 6).

(*d*) *Camouflage detection film.* This film was designed specifically to detect green colored artificial camouflage by recording it as blue to blue-green in contrast to a red recording of natural vegetation. It combines the advantages of both infrared and color films. The structure of the film is such that high infrared reflective objects—natural vegetation—record as red; low infrared reflective objects record as blue or green.

b. Radar. Radar detecting devices emit radio signals, usually in the form of pulses of an ultrahigh frequency, which are reflected from the object being viewed and received back at the point of transmission. By analyzing these

Figure 3. Low oblique photograph.

minute reflected signals, the characteristics of the object under observation may be determined. Concealment from radar depends upon the reduction of this reflection. This can be accomplished by digging in or by use of a defilade. It is important to remember that foliage cover alone cannot be relied upon completely to defeat detection by radar.

c. Infrared. There are two types of infrared detectors: Active (near) which requires illumination of the target by some light source, such as infrared spot or floodlights or the sun; and passive (far) which detects the heat emitted by the target and converts the signal to a visual picture graph or sound record. Concealment from active infrared depends on the reduction of reflectance contract between the object and its surroundings. If the object and its background are of the same reflectance and texture, total concealment is achieved. Concealment from passive infrared is dependent upon reducing heat emission of objects which are hotter than their surroundings. Therefore, some insulation or shield must be used. Defilade, heavy brush, or even tree cover will at least attenuate the heat radiation. To what extent, depends upon the density or thickness.

8. Factors of Recognition

Regardless of the method of observation employed, there are certain factors which must always be present to help the eye and brain identify an object. These are termed factors of recognition (fig. 8).

Figure 4. Black and white film.

a. Position. An object is often identified by its position with relation to its surroundings. A long object on a railroad track is assumed to be a train; similar objects on a river and parallel to its banks are assumed to be boats or barges. A large structure in a group of frame buildings might be a barn. Position is nothing more than the relative space relationship of one object to another object or objects.

b. Shape. Experience teaches people to associate an object with its shape or outline. At a distance, the outline of objects can be recognized long before the details of makeup can be determined. Trucks, guns, tanks, and other common military items of equipment all have distinctive outlines that help to identify them.

c. Shadow. Shadow may be even more revealing than the object itself. This is particularly true when viewed from the air. Such items as factory chimneys, utility poles, vehicles, bridges, and tents have distinctive shadows. Sometimes it may be more important to break up or disrupt the shadow of an object than it is to conceal the object itself.

d. Texture. Texture refers to the ability of an object to reflect, absorb, and diffuse light. It may be defined as the relative smoothness or roughness of a surface (fig. 9). A rough surface, such as a field of grass, reflects little light and casts many shadows on itself. Consequently it appears very dark to the eye or on a photograph. A smooth surface, such as an airstrip or the roof of a building, reflects more light on an aerial photograph. Thus, an airstrip, even though it might be painted the same color as the surrounding terrain, would show up as a lighter tone on a photograph. The almost total absence of texture results in shine. One of the most revealing breaches of

Figure 5. Color film.

camouflage discipline is shine. This alone can attract attention to a location under enemy observation regardless of the type. Shine is generally associated with the reflection of sunlight from windshields, windows, mess kits, and other such almost textureless surfaces. Even the lenses of field glasses, when used in direct sunlight, can reflect a bright shine similar to that of a mirror. (Some substances, such as certain plastics, regardless of the degree of texture, still present a shine.)

e. Color. Color is an aid to an observer when there is contrast between the color of an object and its background. The greater the contrast in color, the more visible the object appears. While color alone will usually not identify an object, it is often an aid in locating the object or confirming a tentative identification. A secondary consideration is the tone of a color. This is the modification of color in varying shades. Usually darker shades of a given color will be less likely to attract an observer's attention than the lighter, more brilliant shades.

f. Movement. The last factor of recognition is movement and although this factor seldom reveals the identity of an object by itself, it is the most important one for revealing existence. Even though the other factors of recognition have been completely eliminated, an enemy observer will be attracted to the area if movement is not controlled. He may even be concentrating his attention on some other area but he will not fail to detect movement in another area through his peripheral vision.

Figure 6. Infrared film.

Figure 7. Camouflage detection film.

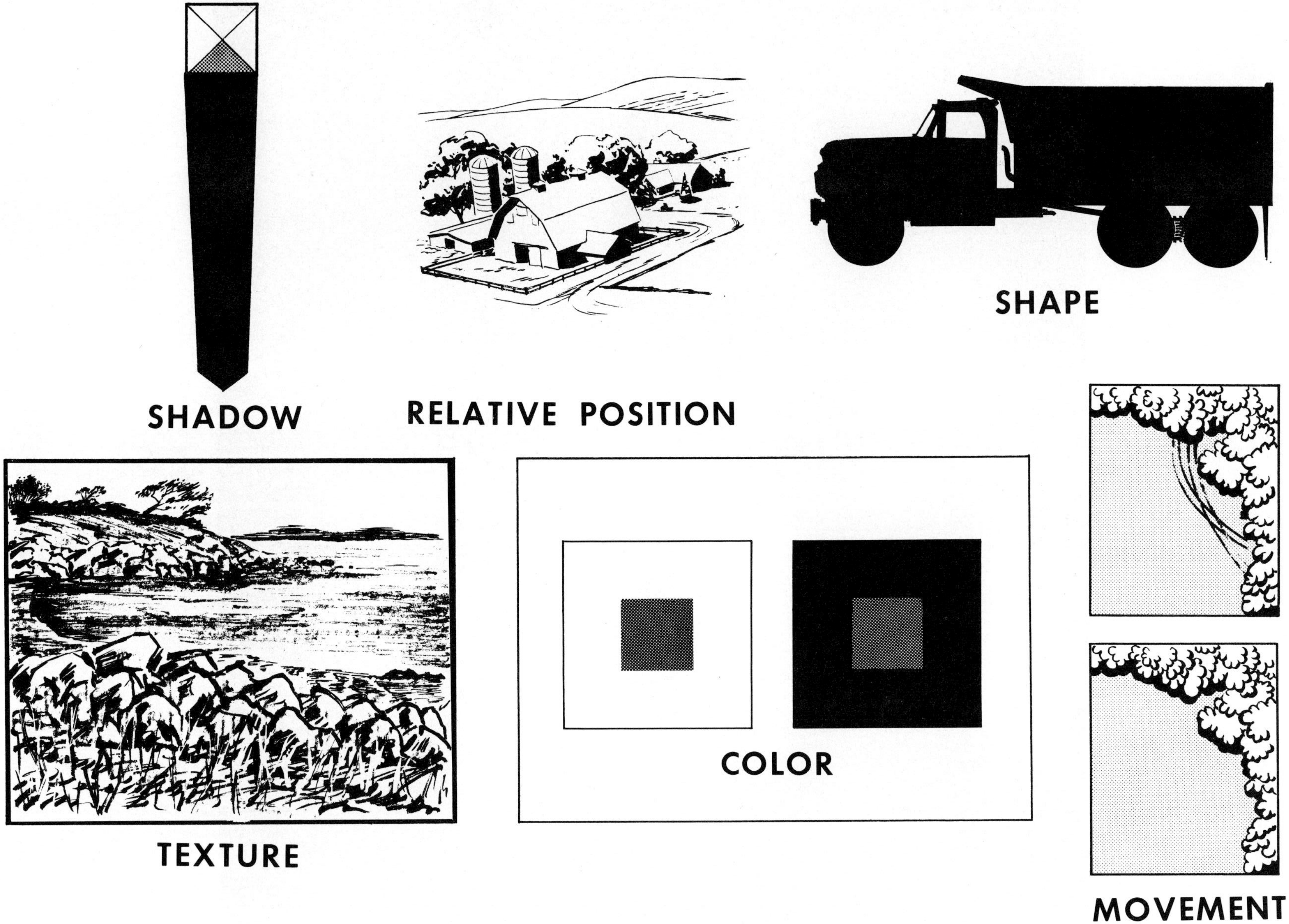

Figure 8. Factors of recognition.

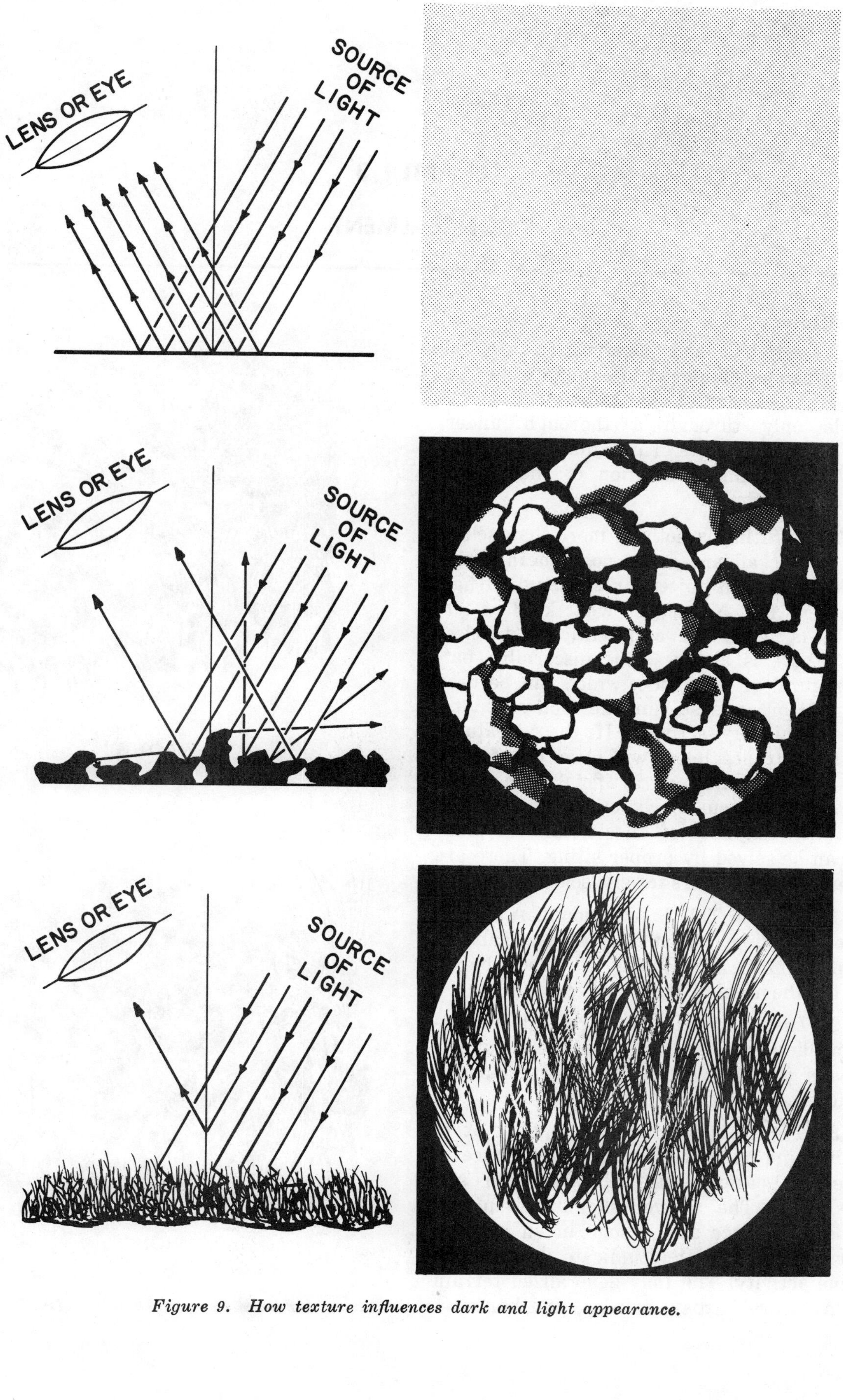

Figure 9. How texture influences dark and light appearance.

CHAPTER 3

CONCEALMENT

9. Principles

Siting, discipline, and construction are the three principles employed to eliminate the factors of recognition. Good camouflage is obtainable only through a thorough understanding of the factors of recognition and careful evaluation and utilization of these principles of concealment.

a. Siting. Siting is nothing more than selecting the most advantageous position in which to hide a man, an object, or an activity (figs. 10 through 14). No matter what kind of terrain it is, even an apparently featureless desert, there is always some discernible pattern, natural or manmade, which can be used to conceal or at least blur the tactically vital signs of military activity. If these features are utilized, concealment will often be effective without employment of artificial camouflage construction measures. Experience has shown that a vast majority of all concealment problems can be solved by proper siting. There are three governing factors for site selection.

(1) *Mission.* This is paramount. A certain location may be excellent from the concealment standpoint, but if it makes it impossible to carry out the mission, it is pointless.

(2) *Dispersion.* The requirement for dispersion dictates the size of the site. A site is useless if it will not permit enough dispersal for effective operation.

(3) *Terrain patterns.* The final point to keep in mind is "What, if any, disturbance in the terrain pattern will this particular site necessitate?" The answer should be "none." This is vital since any change in an existing pattern will immediately indicate the presence of some activity. The four generalized terrain patterns—rural, urban, wooded, and barren—

1 WRONG

2 RIGHT

Figure 10. Choice of position.

Figure 11. Example of proper siting and dispersal of tents in sparsely vegetated terrain (barren).

all have distinctive characteristics, which are necessary to preserve.

(*a*) *Rural terrain.* This terrain is recognized from the air by virtue of its peculiar checkerboard pattern. This is a result of the different types of crops and vegetation found on most farms.

(*b*) *Urban terrain.* A picture from the air of most urban terrains is characterized by more or less uniform rows of housing, interwoven with streets, and punctuated by carefully located trees and shrubs.

(*c*) *Wooded terrain.* The picture that this terrain presents to the aerial observer is a natural, irregular work of nature, unlike the almost geometric pattern of the manmade rural and urban terrain.

(*d*) *Barren terrain.* Like the wooded terrain, barren terrain present the aerial observer with an uneven, irregular work of nature, without the abrupt patterns of the rural and urban areas.

b. Camouflage Discipline.

(1) The second basic condition for the achievement of success in any camouflage effort is the strict maintenance of camouflage discipline, by both the unit as a whole and the individual soldier. This means avoidance of any activity that changes the appearance of an area or reveals the presence of military equipment. It is a continuous, round-the-clock necessity and applies to every individual. If the rigid routine of such discipline, both visual and audio, is not followed by only one man, the entire camouflage effort will fail. Carelessness and laxness will undoubtedly reveal the position to the enemy. Tracks, spoil, and debris are the most common signs of military activity which indicate concealed objects. Therefore, new tracks should follow existing paths, roads, fences, or natural lines in the terrain pattern. Exposed routes do not end at a position, but are extended to another logical termination. If practicable, exposed tracks are camouflaged by brushing out or covering. Spoil and debris are covered or placed to blend with the surroundings (figs. 15 through 17). A camouflage SOP listing rules like the ones mentioned will help a great deal in enforcing camouflage discipline. It should assign to certain individuals the responsibility of enforcing this discipline. It should list rules for conduct of individuals, units, vehicle drivers, etc., in combat, in bivouacs, or in any other situation which may be appropriate to the unit.

(2) No less important is the strict observance of blackout rules. At night, windows, hatches, entrances, and other openings through which light can shine must be covered with shutters, screens, curtains, and other special opaque materials to prevent enemy ground and air observers from noticing the interior illumination. Fires can be lighted only in specially designated and equipped areas. Smoking is forbidden near the enemy, as is the display of

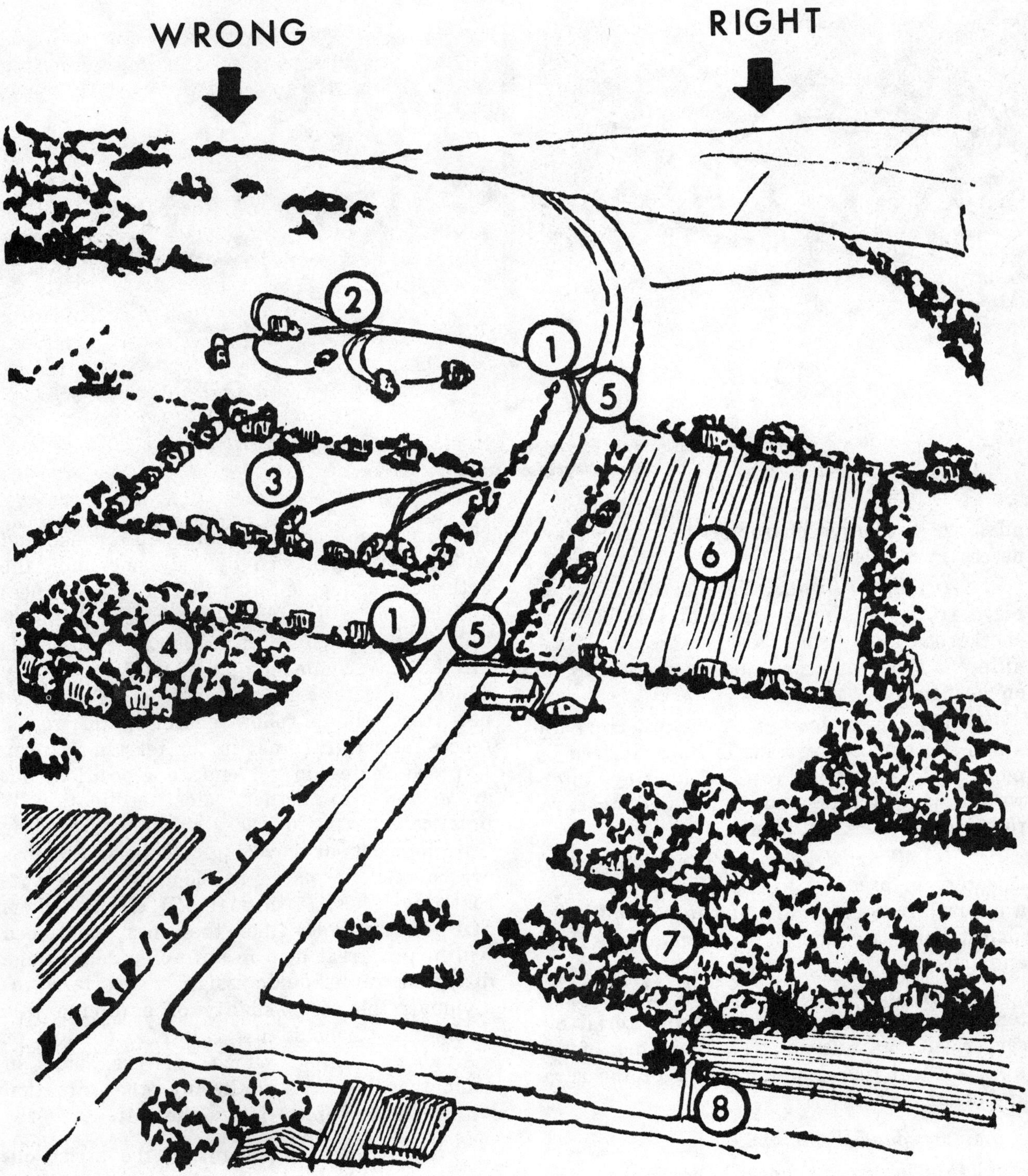

1 Corners cut short
2 Tracks and parked vehicles contrast with terrain pattern
3 Insufficient dispersion and exposed tracks of three vehicles
4 Insufficient dispersion; newly made tracks point to position
5 Existing tracks used for access
6 Correct dispersion. Good use of lines in terrain pattern
7 Correct dispersion and good use of overhead cover
8 Inconspicuous tracks to cultivated field

Figure 12. Use of terrain and proper dispersion.

Figure 13. In regular urban terrain, military objects must be sited parallel to and close to pattern lines.

lights of any types. Combat and transport vehicles can be allowed to travel only with their lights turned off or obscured. Aerial photographs taken at night by the light of flares or by the use of image intensification equipment can pick up breaches in camouflage discipline, which are more likely to occur at night than in the daylight hours. Consequently, *the same standard of camouflage discipline must be adhered to by night as by day.*

(3) Troops must pay special attention to sound camouflage during night movement and apply all the principles of scouting and patrolling. During nighttime river crossings, the noise from the paddles should be muffled. Revealing sounds from tank and truck movement from engineering work can be muffled by stronger sounds, e.g., so called sound screens, created by the firing of machineguns and artillery pieces; the running of tanks, prime movers, and tractor engines along a broad front; or by the operation of sound projection stations which can imitate various battle sounds.

c. Camouflage Construction. The third and final principle on which good camouflage is based is camouflage construction. When the terrain and natural vegetation are such that natural concealment is not possible, artificial camouflage is added. Artificial or natural materials are used to help blend the object or individual with the surrounding terrain. Camouflage construction should be resorted to *only* when siting and discipline cannot produce the desired concealment. Natural materials are preferred over artificial materials, since the former resemble the surrounding vegetation. If artificial materials are used, they must be so arranged as to blend with the surroundings to the greatest possible degree. They must be of sturdy construction to withstand varying weather conditions and be constantly checked and maintained. The construction work must be hidden, with the work parties observing the strictest discipline. If possible, all engineering work should be carried out at night, with all traces of the night's activities camouflaged before morning. There should be no disruption of the terrain pattern; no destruction of plants or trampled grasses; nor should there be any new roads or open ditches visible. This is difficult to do, but unless strict discipline is maintained during such construction, there is little point to the camouflage effort.

10. Camouflage Methods

There are three fundamental ways of concealing installations and activities: Hiding, blending, and disguising.

a. Hiding. Hiding is the complete concealment of an object by some form of physical screen. Sod over the mines in a minefield hides the mines; the overhead canopy of trees hides the objects beneath it from aerial observation; a defilade hides objects from ground observation; a net hides objects beneath it; a road screen hides the obstacle behind it. In some cases, the screen itself may be invisible to the enemy. In other instances, while visible to the enemy, a screen conceals the amount and type

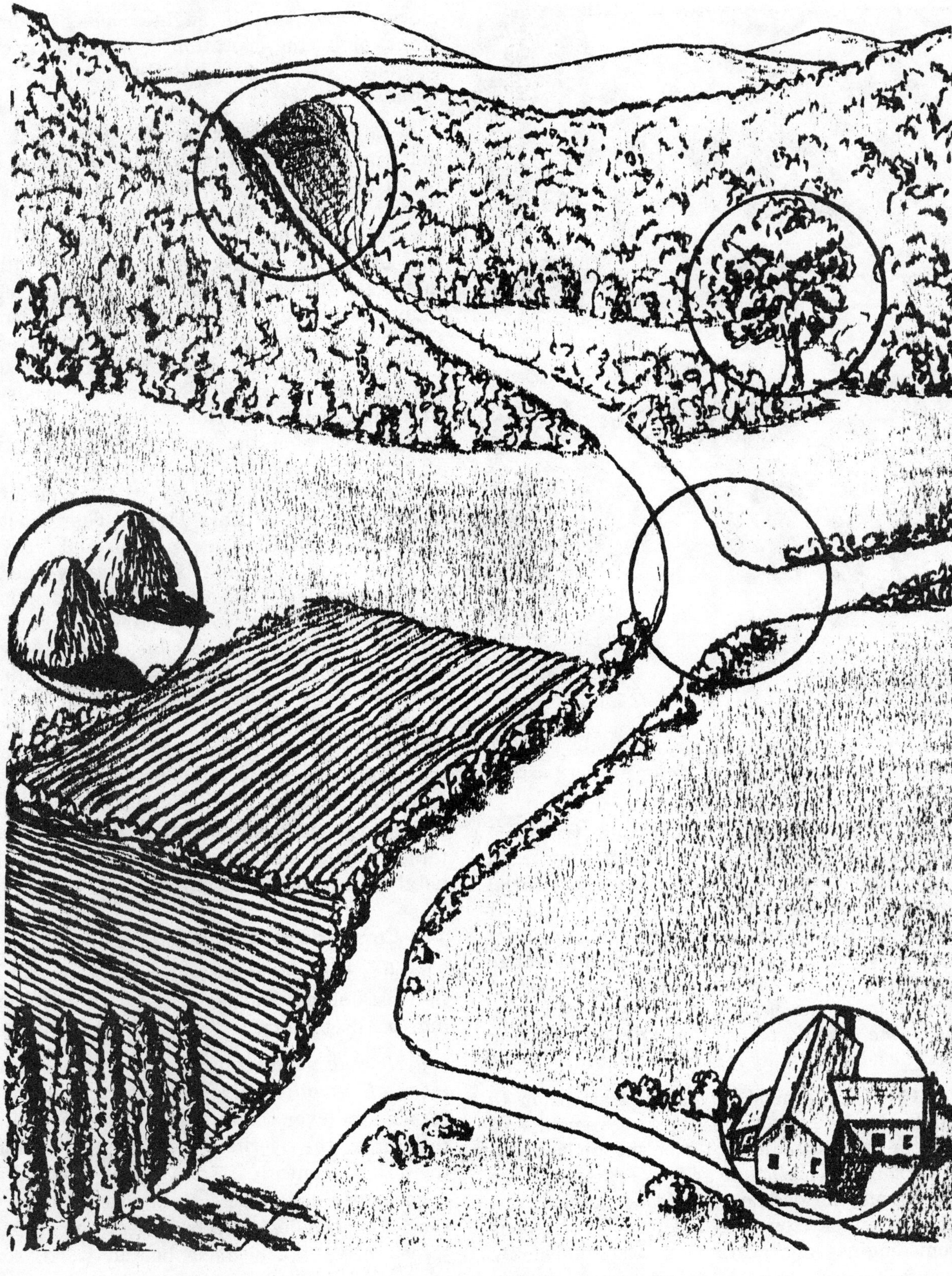

Figure 14. Landmarks attract attention. Position should avoid terrain features which may be used as reference points by enemy ground and aerial fires.

Figure 15. Track discipline.

Figure 16. It is obvious here, to even the untrained observer, that some activity is taking place at both ① and ② and bears watching.

of activity behind it. Figure 18 illustrates two examples of hiding.

b. Blending. Blending is the arrangement or application of camouflage materials on, over, and around the object so that it appears to be part of the background. For example, the individual soldier can apply stick face paint to the exposed areas of the skin; add burlap, paint, and live vegetation to his helmet and clothing so that he will closely resemble or blend into the background. The same things can be done to equipment and structures to make them inconspicuous. Blending distinctive manmade objects into a natural terrain pattern is necessary to maintain a normal and natural pattern (figs. 19 and 20).

c. Disguising. Disguising is the third method. It involves the simulation of an object or activity of military significance. Clever disguises will mislead the enemy as to identity, strength, and intention, and will draw his fire from the real target.

1 Before 2 After

Figure 17. Turnoff tracks are well concealed by clever use of a camouflage net.

It would be dangerous to run this roadblock without first stopping to reconnoiter.

Figure 18. Examples of hiding.

This screen hides from ground observation the amount and type of traffic using this important main supply route in Korea.

Figure 18—Continued.

Figure 19. A tank blending into its background somewhere in Korea.

Figure 20. An installation blended into its background.

CHAPTER 4

CAMOUFLAGE FOR THE INDIVIDUAL

11. Introduction

a. Individual camouflage is the concealment a soldier uses in combat to surprise, deceive, and outwit his enemy. He must know how to use the ground. He must adapt his dress to blend with his background. He must carefully select his routes between positions to gain such concealment as is possible while he is in motion. The simple principles in this chapter have been battle tested. If the soldier learns and practices them continuously in training he will know what to do in actual battle.

b. Individual camouflage activities are designed to deceive two kinds of enemy observers—ground and air. Views from the ground are familiar, but views from the air are usually quite unfamiliar. In modern warfare the enemy puts much reliance on aerial photographs for information as to our activities and intentions. It is important to become familiar with the "bird's-eye-view" of the terrain as well as the ground view in order to learn how to guard against both kinds of observation.

c. Effective concealment of the individual depends primarily on the choice of background and its proper use. Background is the surrounding area seen from the ground and the air. It may be anything—a portion of the jungle, an area in a barren rocky desert, a farm yard, or a city street. It is the controlling element in individual camouflage and governs every concealment measure. The clothes that are worn must blend with the predominant color of the background. Skin and light colored equipment are toned down for the same purpose. The individual soldier must practice blending with the background by hiding in shadows and avoiding contrast between his silhouette and the background. He must avoid movement which the immobile background will emphasize. To keep the appearance of the background free of signs which point to the presence of military personnel and activities, he must follow hidden routes, and conceal spoil, tracks, equipment, and installations.

12. Disguising the Helmet

The outline of the helmet is one of the striking characteristics of a soldiers equipment, and its curved familiar shape can be easily identified by the enemy. One of the first steps for individual camouflage is the disruption, both of the form of the helmet and the strong straight-lined shadow it casts. There are several ways of doing this (fig. 21). Improvised helmet covers can be made from circular pieces of osnaburg, burlap, or other cloth, 20 inches in diameter. A 1-inch hem is sewn around the edges, a drawstring is pulled through it, and the whole cover is pulled tightly onto the helmet. Discarded sand bags, because of their appropriate size make excellent improvised covers. The sack is tucked up into the helmet and the liner then replaced to hold it firmly in place. The covers, regardless of what they are made, should be painted to break up the solid color. Two-inch slits are cut in the cover to allow for the insertion of foliage or bows. No matter what kind of helmet camouflage is used, it is incomplete if the shadow beneath the helmet is not broken up by arranging a bit of foliage or garnishing so that pieces of it hang over the rim of the helmet. Small irregular pieces of cloth, similarly arranged, will accomplish the same thing.

13. Canvas Equipment

Age and repeated washings will fade canvas equipment. When this occurs it must be darkened with paint, mud, charcoal, or anything else that will reduce the tonal contrast.

(1) Rubber bands, or expedient bands made from old inner tubes or burlap strips, secure natural materials. (Note position of band.)

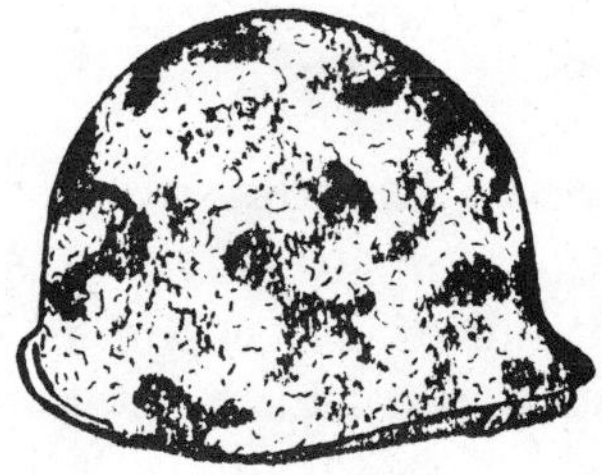

(4) Texturing diminishes shine from steel helmet.

(2) Slits in burlap allow insertion of natural material.

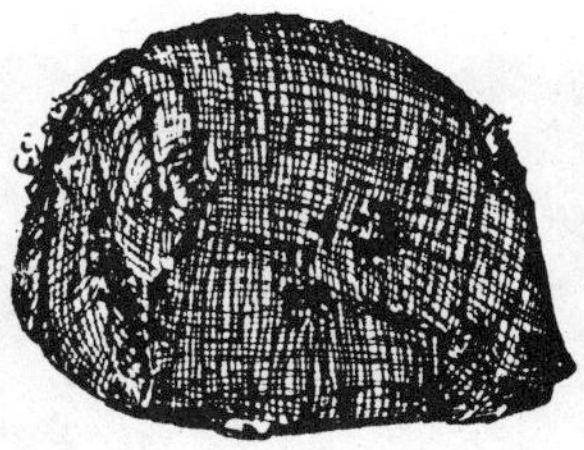

(5) Burlap helmet cover pattern painted to break up solid color before natural materials are inserted.

(3) Form disrupted by burlap bows tied into slitted cover.

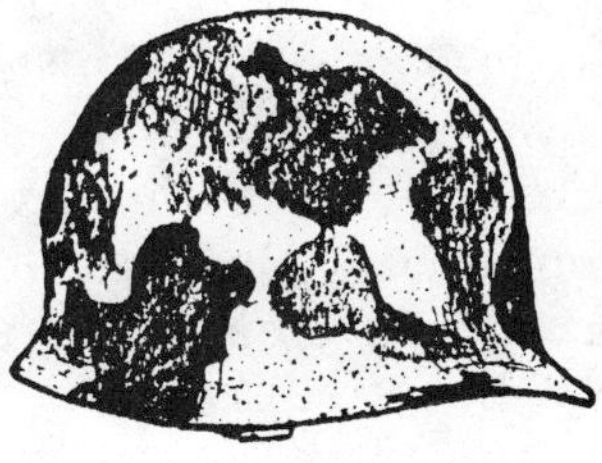

(6) A disruptive paint pattern, with the pattern carried across the curved lines of the edges, especially those seen from the front.

Figure 21. Various techniques for camouflaging the helmet.

14. Skin

Face, neck, and hands should be toned down by painting them in a disruptive pattern or by toning them down in an even color (fig. 22). When using disruptive painting the patterns should cut across the nose lines, cheek bones, eye sockets, and chin lines. Burnt cork and mud can be used, or in the absence of natural materials, the face paint stick may be used. A mesh mosquito netting, properly toned down, is an effective way of breaking up the outlines of the face.

15. Weapon Tonedown

One of the simplest ways to distort the giveaway outline of a weapon is by wrapping it with burlap garnishing or strips of cloth dyed to match the background. Pattern painting the weapon is another excellent method of distorting the weapon outline. The shiny parts can be covered by cloth, paint, or mud. Care must be taken when camouflaging a weapon not to cause interference in the sighting and firing of it (fig. 23).

16. Shiny Objects

Reflection from brightly shining objects is a common breach of camouflage discipline. All shiny objects must be concealed. This includes such items as watches, rings, belt buckles, and mess kit items. A common breach of discipline is the wearing of goggles on the helmet. This is a violation which should be avoided.

17. Camouflage Clothing

Individual concealment requires a little planning and thought and ingenious use of materials at hand. This applies to the camouflage of clothing also. In the absence of issued camouflage uniforms, the soldier can make his own camouflage suit, adapting its color and pattern to the terrain background. Any color-

Figure 22. Shiny and bright skin must be toned down.

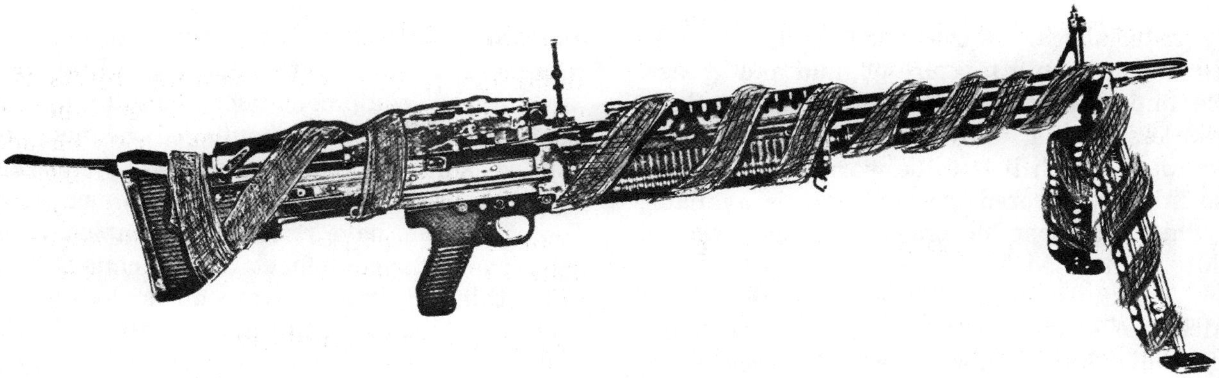

M-60 MACHINE GUN WRAPPED WITH CLOTH

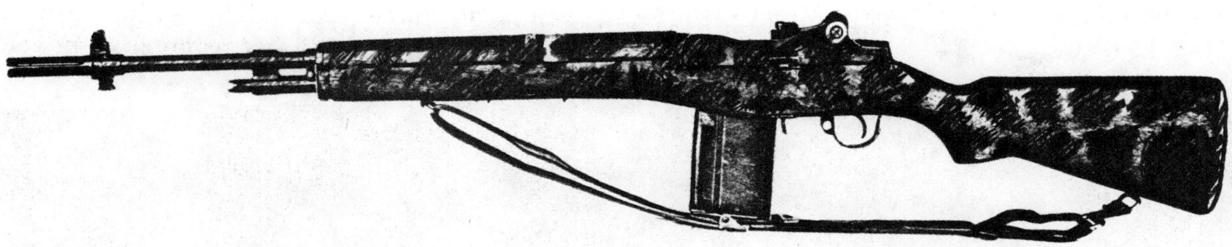

RIFLE PATTERN PAINTED

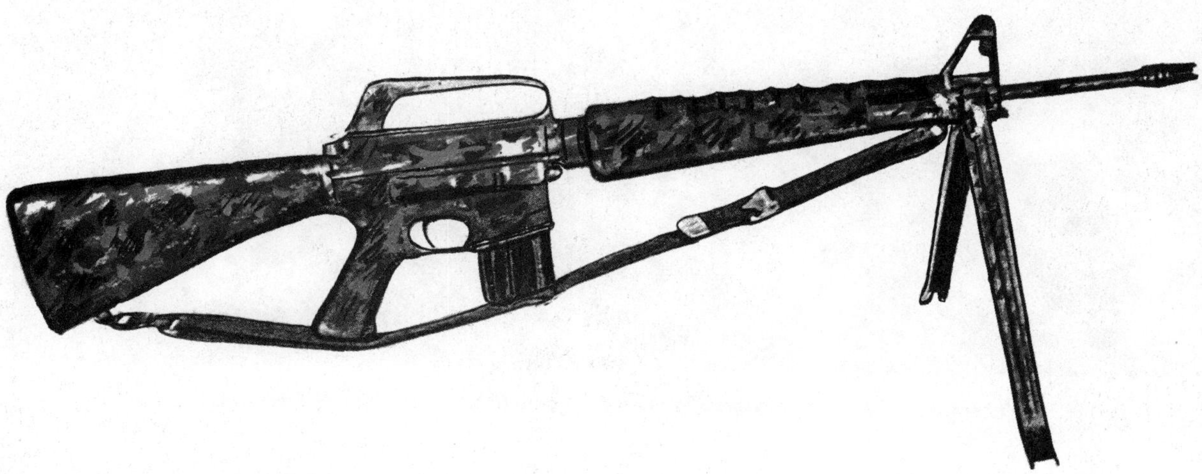

AR 15 - COVER SHINE WITH MUD

Figure 23. Suggestions for camouflaging an individual weapon.

ing materials can be used, such as dyes, crank-case oil, or even a mixture of mud and grease. The important thing is to make the clothing look less like a uniform and more like the terrain in which it is to be worn (fig. 24). For use in snow covered terrain there is available a white garment designed to blend with a white or mottled white and black background. The snowsuit does not conceal the small patches of shadow that surround a human figure, but this is not necessary since snow country is seldom all white—it does contain numerous dark spots and shadows. If certain snow areas are all white with absolutely no shadows, use is made of defiles, snow drifts, and natural folds in the ground. It must be remembered that camouflage clothing and camouflaged equipment alone won't conceal. The tonedown, the camouflaged helmet, the painted suit, the covered shiny objects are just the beginnings of the concealment job. Too often, men have relied with complete faith on a camouflaged helmet and a camouflaged suit, thinking themselves miraculously invisible, and completely disregarding all the basic elements of camouflage. This invariably leads to disastrous results. These measures make it easier to conceal, but only when taken in conjunction with the principles of scouting and patrolling (FM 21–75).

Figure 24. Self-made camouflage suits blend this patrol into the terrain.

CHAPTER 5

FIELD FORTIFICATIONS

18. Siting

a. After the demands of the military situation and the mission have been met, siting with proper background is the first consideration given to the concealment of a fortification. From the standpoint of ground observation, the emplacement should be sited to avoid creating a silhouette against the sky or against a background of contrasting color. To avoid air observation, the emplacement should be located under trees, bushes, or in dark areas of the terrain.

b. It is equally important that the concealing cover chosen is not isolated, since a lone

Figure 25. Spoil clearly reveals the positions.

clump of vegetation or solitary structure is a conspicuous hiding place and will draw enemy fire whether the enemy sees anything or not.

c. The natural look of the terrain should not be disturbed. This is best accomplished by removing or camouflaging the spoil (fig. 25).

d. Natural terrain lines, such as edges of fields, fences, hedge-rows, and rural cultivation patterns are excellent sites for emplacements to reduce the possibility of aerial observation. Regular geometric layouts are to be avoided.

19. Construction

Before any excavation is started, all natural materials, such as turf, leaves, forest humus, or snow are removed and placed aside to be used later for restoring the natural appearance of the terrain. Concealment while constructing an elaborate fortification is vital.

20. Covers

When a position cannot be sited under natural cover, camouflaged covers are valuable aids in preventing detection (figs. 26 through 28). Materials native to the area are preferred, but when using natural materials over an emplacement they must be replaced before they wilt and change color, leading to detection. Artificial materials may be used effectively, such as those made to simulate tall grass, bushes, stumps, and rocks, whichever the terrain calls for. They are valuable principally against aerial observation. They are light in weight and may be easily pushed out of the way.

Figure 27. Details of a foxhole cover.

Figure 26. Cover made from burlap and tree bark.

21. Machinegun Positions

The machinegun receives the close attention of enemy troops and its concealment must be as perfect as possible. Usually, machinegun emplacements are hasty, in which case camouflage means siting to best advantage and then using any materials at hand.

Figure 28. A roadside foxhole with a 3.5 rocket launcher. (Back blast from such recoilless weapons as the rocket launcher must be considered when locating and constructing emplacement.)

22. Mortars

Mortars should always be sited in defilade. Since a mortar covering a designated target area has a wider choice of position than the other smaller weapons, such defilade can almost always be found and concealment from direct ground observation is fairly easy. Proper siting in shadow and broken ground pattern making certain there is the necessary overhead clearance for firing, together with intelligent use of natural and artificial materials offer the required mortar concealment from the air.

Figure 29. Mortar emplacements.

CHAPTER 6

VEHICLES AND ARTILLERY

Section I. VEHICLES

23. Revealing Factors

A badly concealed vehicle can lead to much more than just a lost vehicle; it may mean discovery of a unit, disclosure of an important tactical plan, or complete destruction of an installation. Camouflage of vehicles depends not only on concealing the vehicles themselves, their shine, shadow, and shape, but equally on preventing and concealing their all-revealing tracks. (It should not be forgotten that enemy ground and aerial observation is drawn quickest by anything that moves, and that nothing can be done to conceal vehicles moving through undergrowth or along exposed routes.)

a. Tracks. Tracks are especially revealing to the aerial observer. They indicate type, location, strength, and even intentions of a unit. The gradual turns of wheeled vehicles are distinguishable from the skidding turns of a track laying vehicle, and often a single track across an area of low vegetation is clearly visible. The last is especially true in the early morning hours when there is a heavy dew. Tracks should follow closely and be parallel to hedges, fences, cultivated fields, and other natural terrain lines in order to remain inconspicuous from the air. Also, tracks should always continue past the position to a logical termination.

(1) Completely concealed roads rarely exist. Even the comparatively small amount of timber which must be cut down to clear a roadway through a wooded area leaves gaps in the overhead cover that are clearly seen from the air. Partially concealed roads do exist however, and they are better than exposed ones. Reconnaissance parties should locate them. Any gaps in overhead cover on such a road can be concealed by erecting overhead screens or either artificial or natural materials. On short stretches, exposed tracks may be erased by brushing leaves and debris over them. All concealed routes should be marked in advance of use and guards posted to insure minimum disturbance in the area.

(2) If possible, before a unit occupies a position, a quartering party should first make a reconnaissance and lay out a concealed track-plan. No vehicle should enter the new area until then. This plan should be laid out to fit into the terrain pattern as inconspicuously as possible by taking advantage of existing roads, overhead cover, and shadow casting lines. Many factors must be considered in such a plan: duration of occupation; time allowed for entering and leaving; size, character, and mission of occupying unit; distance from the enemy; and weather effect on visibility. A standard track plan is impossible—an individual solution is required for each installation. In addition to laying out a plan on the ground itself, a plan should be sketched on either a map overlay or a sketch of the area. Parking areas should be indicated as well as those portions of routes to be patrolled by traffic guides.

(3) Since a unit may have to occupy a position without prior reconnaissance, unit camouflage training must insure that all personnel are trained to follow terrain patterns and utilize all overhead cover, when possible. Particular attention must be directed to training of vehicle drivers, so that they will follow these rules automatically, even in the absence of NCO's and officers. The officers and NCO's must instruct all personnel that when the first vehicle enters an area, guards must be sta-

tioned at critical points to direct traffic. This prevents unnecessary vehicle slow-down, stopping, or jamming on a roadway.

(4) In snow covered terrain, concealment of tracks in a major problem. Even in light snow, tracks make strong shadow lines visible from great distances. Sharp turns should be avoided because the resulting snow ridges cast even heavier shadows. The same principles stressed throughout this discussion apply to snow covered terrian, with a bit more emphasis on following natural shadow casting terrain lines. It is also important that all vehicles keep to the same tracks. Vehicles leaving the track or road may achieve short periods of track concealment by driving directly into or away from the sun, as shadows cast by these tracks will not be apparent until the sun strikes them from an angle. Short lengths of tracks may be obliterated if they are not too deep, by trampling them with snowshoes.

b. Shine. Siting and track discipline do much to conceal a vehicle, but shine can nullify the best site and finest track discipline. Shine is always present when there is light in the sky, sunlight, moonlight, or the light of flares. It is caused by windshield, headlight, cab window, wet vehicle body, and even by the light paint of the insignia. These danger spots must be concealed by any means. The betraying nature of shine should never be underestimated. Even under heavy overhead cover, shiny objects may be revealed through the smallest of gaps.

c. Shadow. There are two kinds of shadows to consider in camouflage. One is the concealing shadow cast by objects on the ground. From the air, these appear so dark that a vehicle parked within them has a good chance of remaining undetected (fig. 30). In the northern hemisphere, the north side of an object higher than the vehicle is the best side on which to park; the east and west sides are dangerous for half a day. The other kind of shadow to consider is that cast by the vehicle itself. This revealing shadow must be hidden by parking either in the shadow of a larger object as explained above, or by parking on the sunny side of the object (fig. 31). In addition, the smaller shadow areas contained within the vehicle itself such as the shadow line of the truck body in and around the cab, beneath the fenders, within the wheels, and in the open back of the cargo space must be blocked out for they too facilitate identification. In snowy areas with little or no cover, vehicles can be parked facing directly into the sun to reduce the shadow, which can then be further reduced and broken up by large snowballs or deep holes dug in the snow. Snow thrown on the wheels helps to disrupt this tell-tale area.

24. Camouflage Measures

a. Siting and Dispersion. As is always the case in camouflage, the aim of good vehicle siting is to occupy the terrain without altering its appearance. To do this, vehicles should be parked under natural cover whenever available. When cover is inadequate, they should be parked so that the shape of the vehicle will disappear into the surroundings. Before a driver can site his vehicle to take advantage of the concealment possibilities of his surroundings, he must know how the different terrains look from the air. In combat zones this knowledge is as important as knowing how to drive the vehicles.

b. Use of Natural Materials. While good siting and dispersion are essential, sometimes they are not enough. Greater concealment can be had by supplementing these measures with natural materials to break up the shape and shadows of the vehicles. This material is almost always available near a parking site and can be erected and removed quickly. When cut foliage is used, it should be replaced as soon as it starts to wither (figs. 32 and 33). Altering the color of vehicles or adding texture to them are other ways to supplement siting and dispersion. Color may be changed by applying mud to the body and tarpaulin, following the patterning principles given below. Texture may be added all over or in pattern shapes by attaching leaves, heavy grass, or course sand to the surface with an adhesive.

c. Pattern Painting. Pattern painting of a vehicle is not a cureall. It is, however, a valuable supplement to other camouflage measures. Added to good siting, dispersion, discipline, and the use of nets, it increases the benefit to be derived from such measures. Vehicle patterns are designed to disrupt the cube shape of vehicles from all angles of view, to disrupt shadows, and to tie in with the shadow at the rear of a vehicle when it is faced into the sun, as well as the large dark shadow areas of windows, mudguards, wheels, and under-

Figure 30. Use of larger shadows.

Figure 31. Throw the shadow onto something irregular.

carriage. The patterns must be bold enough to be effective at a distance. White or light gray paint is applied to the undersurfaces of the vehicle to cause them to reflect light, thus lightening the dark shadows of the undercarriage. This is termed "countershading." As previously stated, pattern painting alone will not conceal a vehicle. To be effective, it must be combined with proper background and siting. Because today's modern mechanized and highly mobile units have the capability of traveling great distances over varied terrain and growth, pattern painting effective one day may be totally ineffective the next day. In fact, the pattern may even prove to be a detriment by rendering the vehicle conspicuous. In areas where snow is a daily problem, vehicle concealment is made much easier if the vehicle is painted with the snow pattern shown in figure 34. The national symbols have been left off the vehicle in the following illustrations in order to show the pattern more clearly. Whether or not to eliminate them and other common vehicular markings must be determined by higher authority.

d. Nets. The principal artificial materials used to conceal vehicles are drape nets. They are easy to use, quickly erected, and quickly removed. Drape nets can give complete concealment against direct observation, but, as with most artificial camouflage materials, can frequently be detected by photographic observation because they often fail to blend properly with the background. In every case, however, properly suspended and supported, drapes do conceal the identity of a vehicle, even though the drape net itself may be detected. Nets are not recommended in snow areas. They require excessive maintenance, cannot support a snow load, and become wet, frozen, bulky, and hard to handle.

e. Digging In. In a desert, or any open barren terrain, the lower an object is to the ground, the smaller is its shadow and the easier it is to conceal from aerial observation. When the situation permits, every effort must be made to dig in important vehicles. Not only are they more easily concealed but they are also protected from fragments. An excavation is made, with a slanting approach and the vehicle is parked in the pit. Sandbags are used to form a revetment for protection and the whole thing is covered with a net. The net is sloped gently out to the sides and staked down. Finally, the vehicle tracks to the position are brushed out or covered.

Figure 32. Park a vehicle close to a clump of trees and use cut foliage to break up its shape and shadows.

The characteristic black shadow in the open end of a cargo truck can be seen for a considerable distance. One way to conceal this shadow is to drop the rear tarpaulin, another way is to use natural materials, as shown here.

Figure 33. Further measures to conceal vehicle.

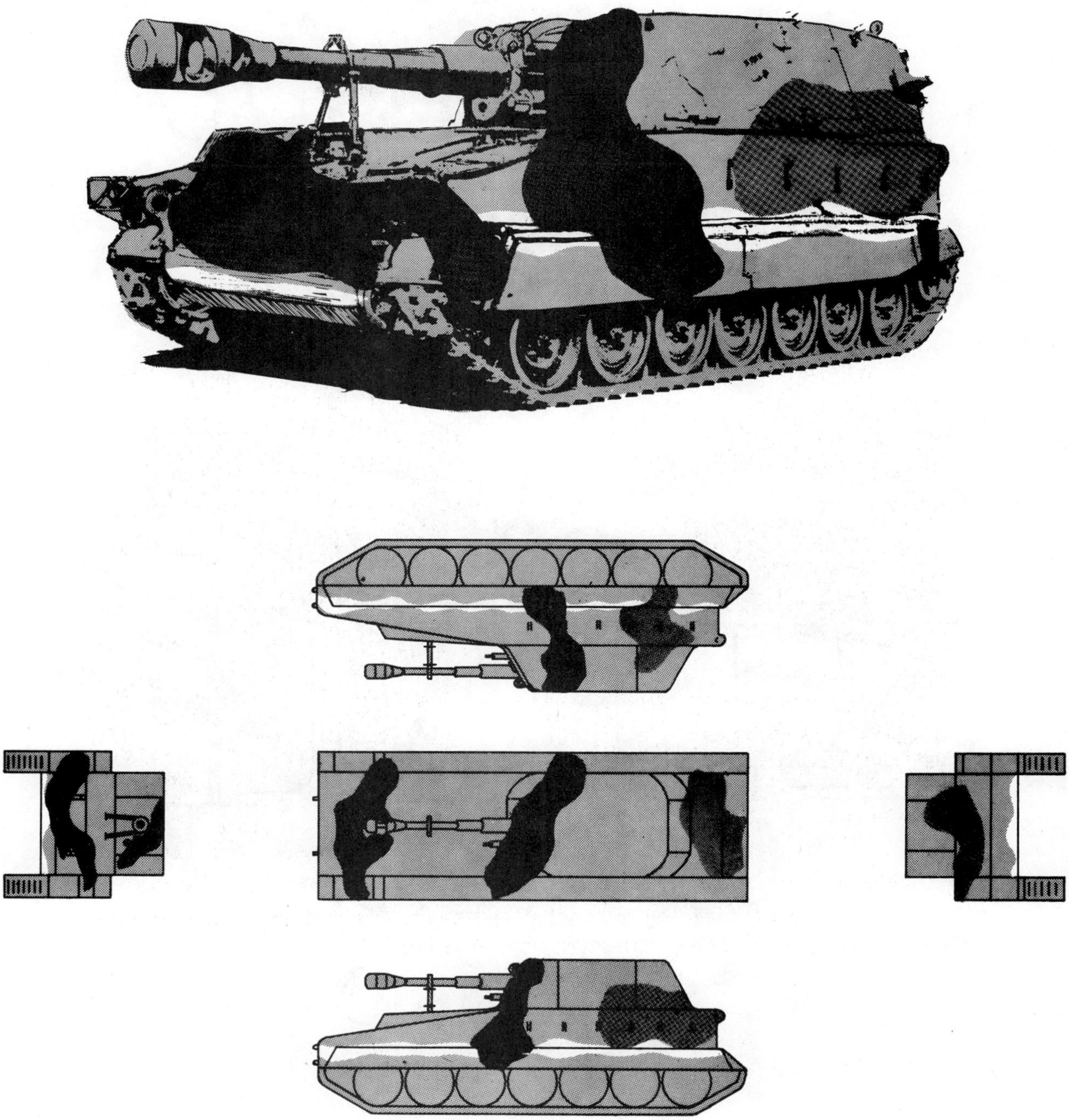

Ground view and pattern plan of tank painted olive drab and black, the under-surfaces countershaded white. Keep patterns bold and simple.

Patterns for temperate zones and jungle

Figure 34. Pattern painting of vehicles.

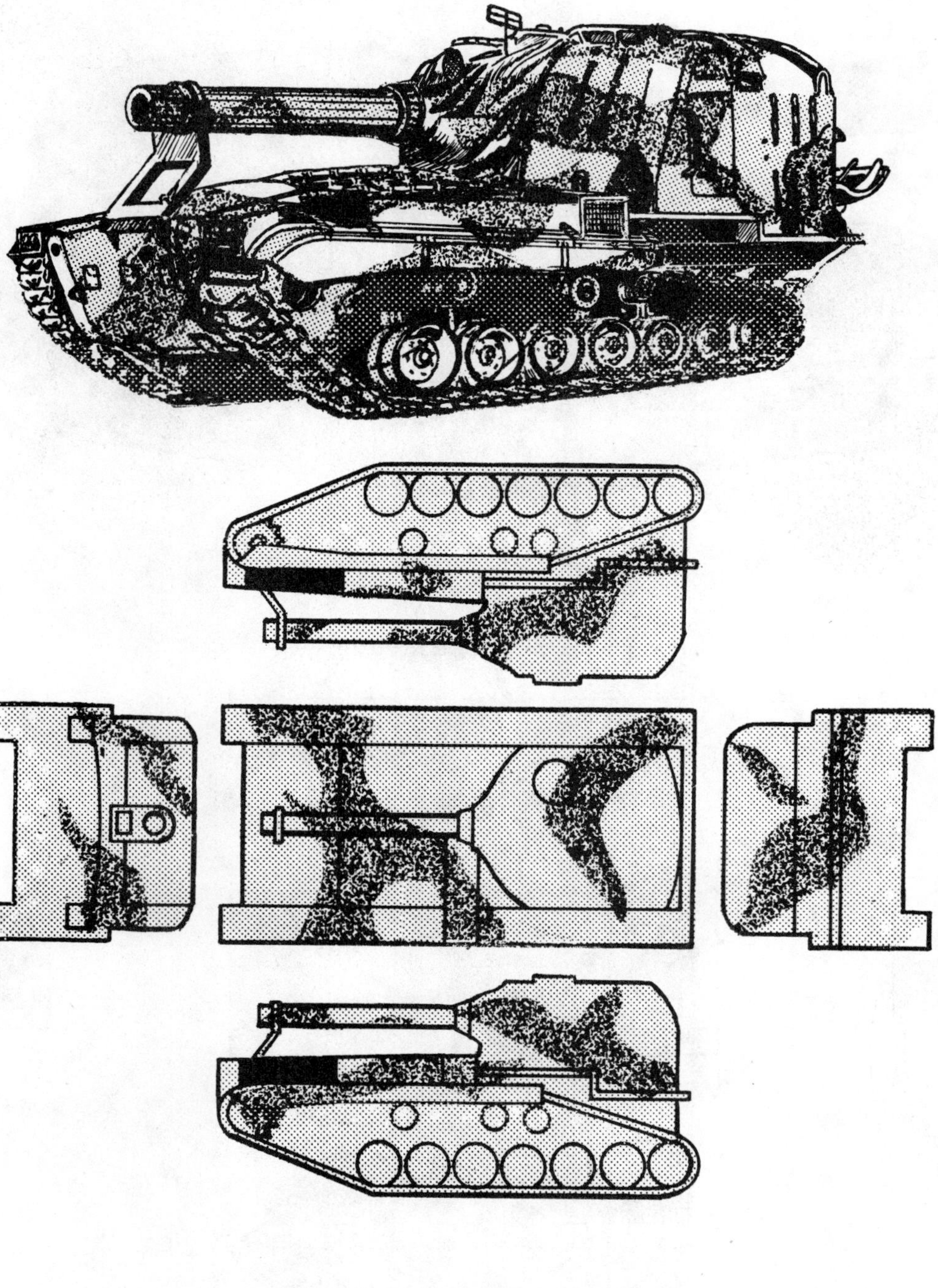

Olive drab and earth red blend with reddish desert backgrounds. Other light colors useful in desert are sand and earth yellow. Patterns break up angular lines.

Pattern for desert terrain

Figure 34—Continued.

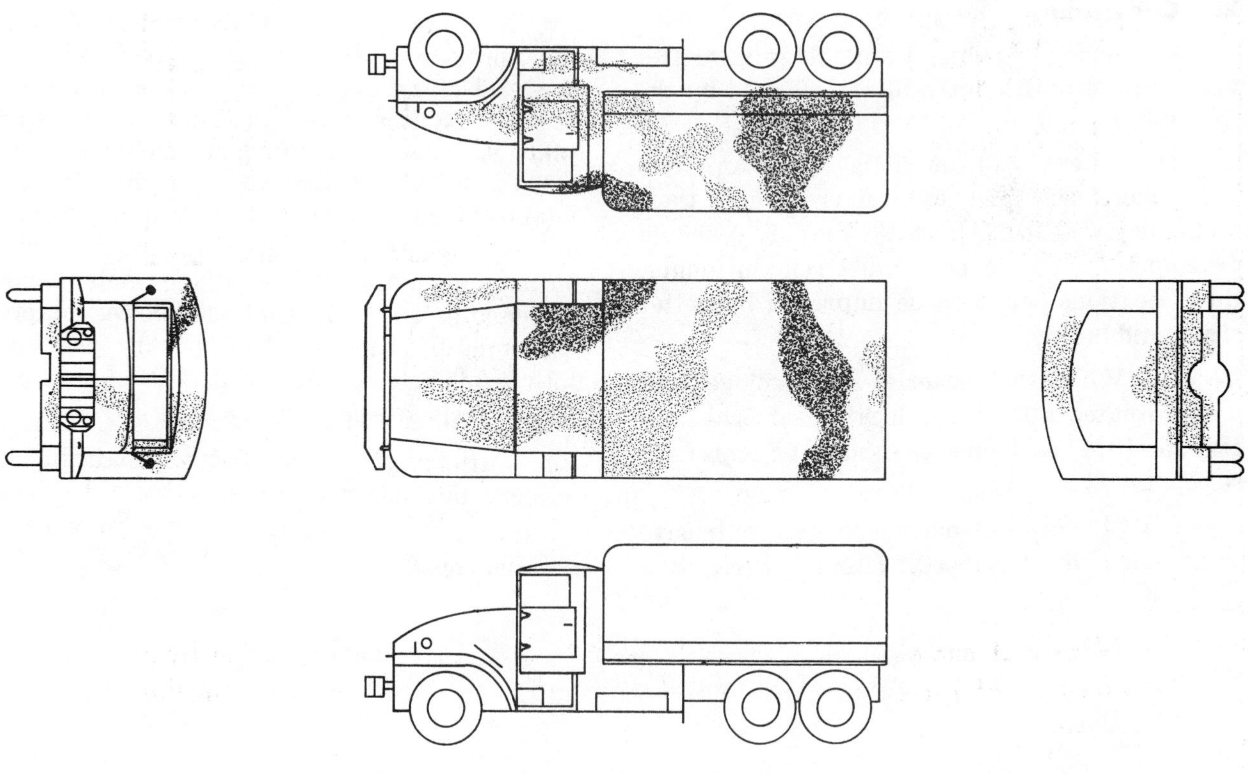

(3) White and olive drab for backgrounds of snow and trees. An equally effective scheme is black and white. In snow, countershading is not necessary.

Pattern for snow terrain

Figure 34—Continued.

25. Revealing Factors

As is the case with all other impedimenta of warfare, skillful concealment of artillery weapons can add immeasurably to the element of surprise and thence to the defeat of the enemy. Enemy observers are trained to search for certain definite signs which indicate the presence of artillery—imperfectly camouflaged weapon positions, blast areas, litter, paths or wheel tracks, and in the case of a missile site, the excessive earthworking scars in the terrain pattern necessitated by a level firing pad and fueling entrances and exits, etc. Even though the weapons themselves are hidden, such signs are dead giveaways of the presence of artillery. These signs may not of themselves indicate the exact nature of the position, but they do attract enemy attention and invite more careful observation.

26. Camouflage Measures

a. Governing Factors. Camouflage measures vary with the situation and are affected by the following:

(1) There will be little opportunity to camouflage positions extensively when their occupancy will be of short duration. If it should develop that the weapons must remain longer, their locations can then be improved by better siting and hiding.

(2) When the batteries are deployed for a coordinated attack, the location of each battery and of each piece should be carefully selected.

(3) In a defensive action, extensive camouflage is developed. Utmost precaution must be taken to deceive the enemy as to the location of the installation.

b. Siting. The exact position for the elements of a battery, within the assigned area, must possess several qualifications:

(1) The required field of fire.

(2) Room for dispersion of weapons, vehicles, and other equipment organic to the battery.

(3) Opportunity to establish communications without creating attention getting ground scars and paths.

(4) Opportunity for access and supply routes. It is desirable to have routes available to the front, flanks, and rear. This is important in situations where it may be necessary to make sudden changes in position. When personnel, ammunition, equipment, and other supplies are moved into position, they must follow a prepared traffic plan (para 23*a*(2)).

c. Nets. Wherever natural concealment is impossible or difficult, suitably garnished twine nets and chicken wire are quick and effective means of concealment. Care must be taken to follow the correct methods in their use (TM 5–200). Wire netting, although heavier and bulkier, holds its form better, is more durable, and is invaluable for positions of a relatively permanent nature. The twine nets, being lighter and easier to handle are better adapted to mobile situations and temporary positions. Both kinds can be garnished with cloth strips and natural materials (figs. 35 and 36).

d. Pattern Painting. Pattern painting of artillery pieces can be an effective aid to concealment and is designed for use in varying terrains (fig. 37).

Honest John missile concealed by use of natural materials

Figure 35. Two ways of concealing missiles.

Net supplementing natural materials for concealing Honest John missile.

Figure 35—Continued.

Figure 36. Net set supplementing sparse vegetation to conceal a rocket launcher and erector.

This Howitzer is painted white for snow terrain in wooded areas, with about 15% left olive drab. In greatly predominant snow areas the piece should be solid white.

This pattern is suitable for light desert backgrounds. In reddish desert background, the earth yel-low should be changed to earth red.

For temperate and jungle terrain, olive drab, field drab, and white.

Figure 37. Patterns for artillery.

CHAPTER 7

AIRCRAFT

27. Introduction

The measures for concealing aircraft on the ground follow the same principles of camouflage as those for the concealment of any vehicle or item of equipment. However, the camouflage of fixed and rotary wing aircraft poses some unique problems which must be considered. These are the effects of propeller and rotor wash; the size and shape of the aircraft; the difficult ground handling characteristics; and the delicate nature of control surfaces, antennas, and aircraft skin.

28. Siting

Aircraft must be situated in an area with easy access to taxiways and runways. This is an operational requirement with which camouflage must not interfere. Aircraft parked in line are extremely vulnerable. Particular care should be taken to select parking sites in which the aircraft may be blended with the natural vegetation, while at the same time observing proper dispersion (fig. 38). Good siting in itself does not eliminate possible detection. Where natural overhead cover is nonexistent, improvisations can be effective (fig. 39).

29. Shine and Insignia

Light reflections from the metal and plexiglass surfaces may be observed shining through even the best of overhead concealment—natural or artificial. The moment an aircraft is parked

Proper dispersion means more than just spacing parked aircraft at a distance from one another as shown above

Figure 38. Aircraft parking.

Military objects must be scattered in a random pattern to escape damage from enemy attack

Figure 38—Continued.

and camouflage begun, all glass, plastic, and shiny metal surfaces must be covered. Extreme care must be used when covering aircraft surfaces. All covers must be secured and it is recommended that some means for marking each cover be used to prevent attempted take-off while such covers and camouflage materials are attached.

30. Supplementing Concealment with Artificial Materials

In many cases it will be necessary to resort to the use of artificial camouflage materials to supplement inadequate natural terrain features and vegetation.

a. Shadow Nets. Shadow nets placed on the ground, under the craft, will break and distort the outline of the shadow, even if the craft is in the open (fig. 40). When such materials are used, they must be tightly secured to well-anchored stakes or other suitable attachments. Otherwise, camouflage materials may be drawn into the rotor blades or propellers of air vehicles. Additionally, care must be exercised when erecting or placing artificial materials, so as to prevent damage to the aircraft and not interfere with operations and maintenance during pre- and post-flight inspections.

b. Hammocks. A twine or wire net hammock hung between trees gives additional overhead protection in sparsely wooded areas. The nets

Where no overhead is available, break up the clear cut outline of the plane's shadow on the ground. Notice how a few freshly cut branches distort the shape and shadow of this aircraft.

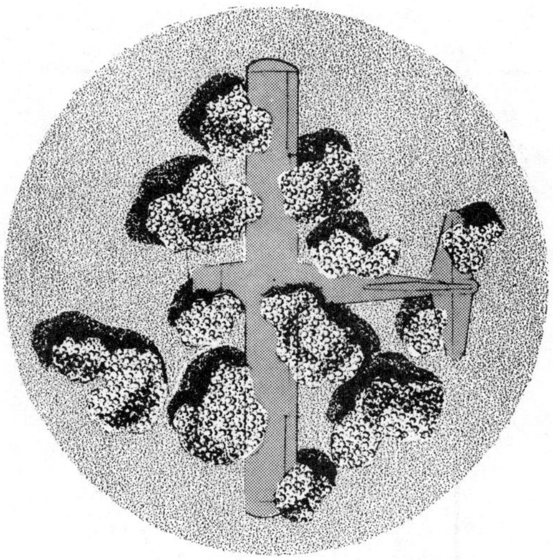

If overhead growth is too thin for adequate concealment, branches can be pulled together with rope, vines, or wire.

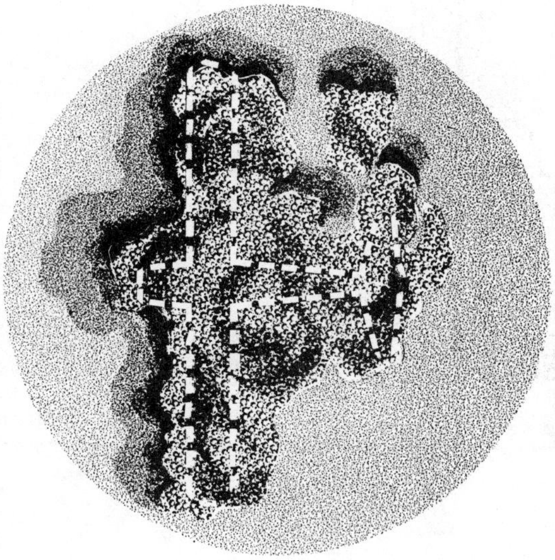

Another method of preventing the tell-tell shadow of a light aircraft is to place the plane in a slight natural depression in the ground. The shadow is still there but so close under the wing surface that it is not easily seen from the air. Place scrub growth on the tail assembly and the nose. The aircraft may escape notice of enemy observers.

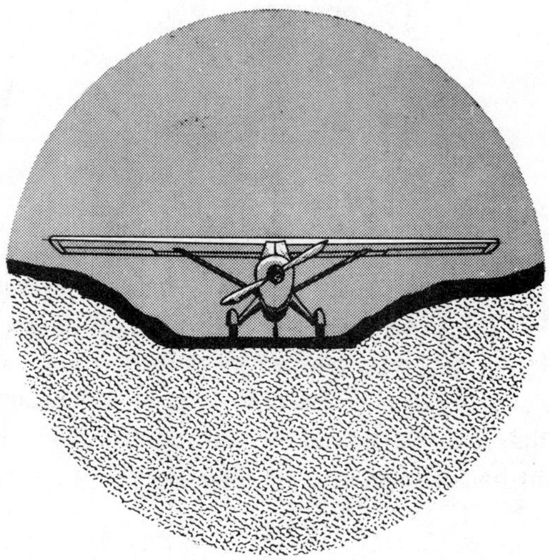

Figure 39. Natural materials for improvising overhead cover.

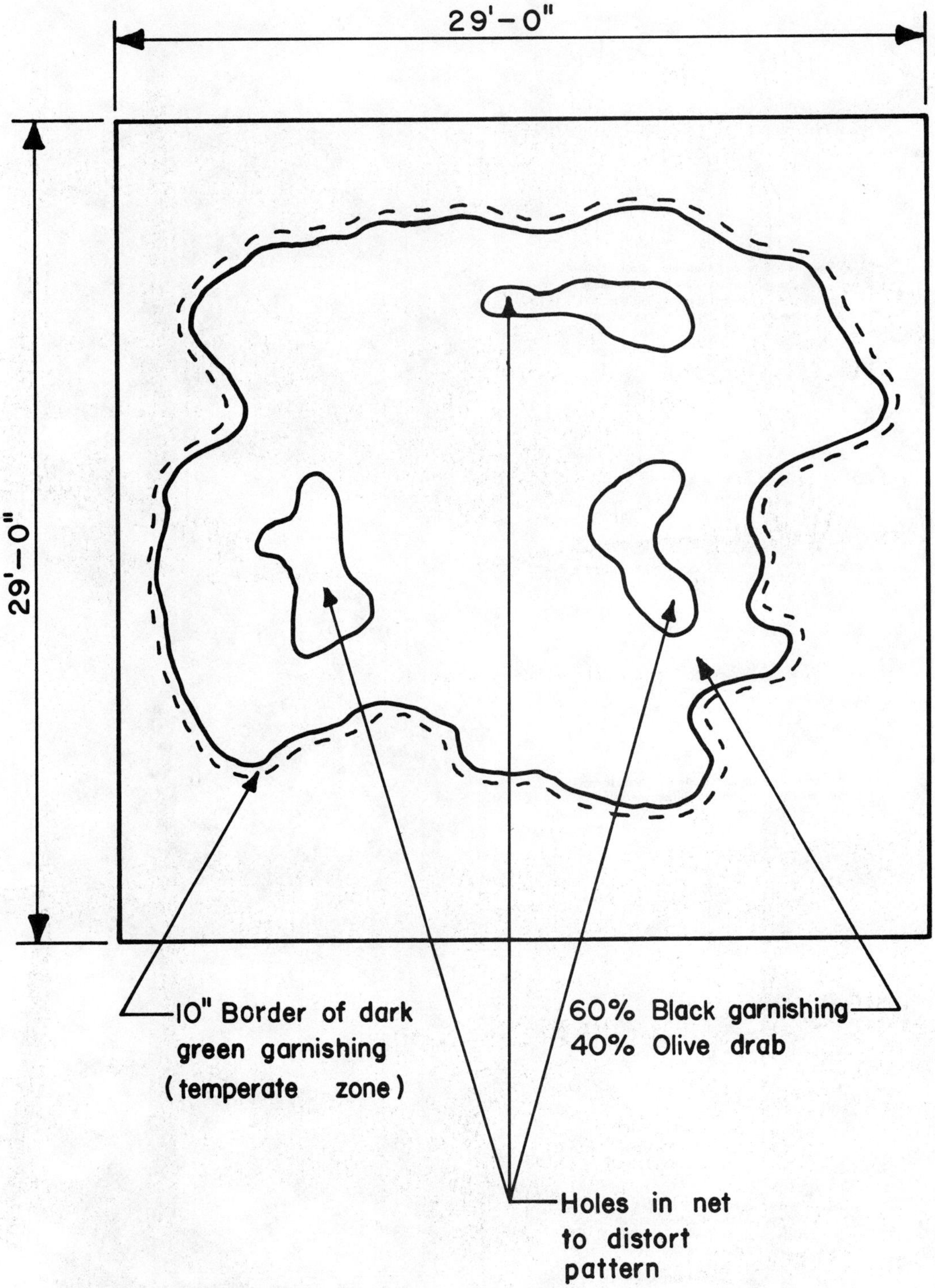

Shadow nets are camouflage nets garnished to appear as a group of irregular dark patches. If the net is not available, dark cloth with holes torn in it can be used successfully, or dark patches can be produced by burning rags, brush, or other debris to produce ashes, or even by pouring crankcase drainings on the ground. Nets and cloth do have the advantage in that they can be reused. If helicopters are to land and take off from these nets or cloth care must be taken to anchor them securely against the wind created by the rotor blades.

Figure 40. Shadow nets.

(1) Hammocks conceal from vertical viewing only.

(2) If additional camouflage is needed, to protect from oblique observation, a garnished twine net can be hung from the exposed sides.

Figure 41. Hammocks used as an overhead cover.

can be garnished with either artificial or natural materials, or a combination of both (fig. 41).

c. *Umbrella Screens.* Another technique is to suspend a series of overhead frames or umbrellas from wires strung between trees (fig. 42). Devices of this type are suitable for small aircraft only.

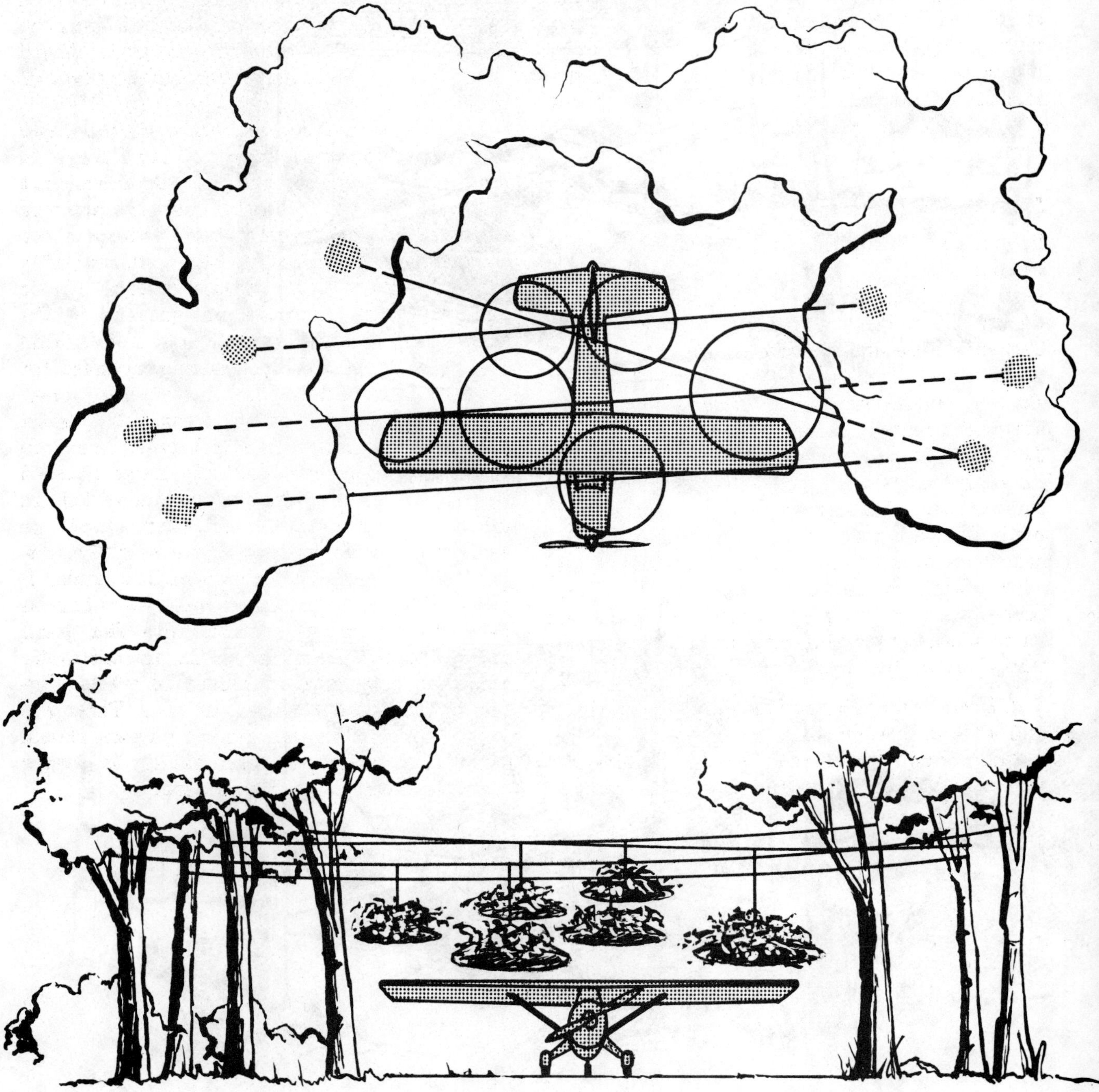

Hung at varying heights above the aircraft, these umbrellas simulate tree tops. They allow quick getaway and servicing. Garnishing may consist of burlap, steel wool, glass fiber, chicken feathers, or fresh foliage (the latter must be maintained and restored as it whithers).

Figure 42. Umbrella screens.

d. Drapes. To conceal completely against direct enemy observation, drapes must be located near trees or bushes, or on ground broken by low scrub growth. If the terrain is bare of vegetation, siting should be in the shadow side of folds in the ground or within mottled ground patterns, such as are formed by rock outcroppings. In such terrain, drape nets will hide the identity of aircraft but will not conceal them completely from enemy aerial photo interpretations.

31. Pattern Painting

Proper painting helps an aircraft blend with its surroundings by distorting characteristic features, minimizing color contrast, and reducing shine. This is all it will do. While these elements of recognition are the revealing ones, it is more often that the aircraft's shape, shadow, and accompanying evidence of activity are the primary causes of enemy detection. The value of pattern painting must, therefore, be regarded as an asset only, when used in conjunction with camouflage principles and methods, i.e., good siting, dispersion, track-concealment, discipline, and the intelligent use of artificial and natural materials. It follows that pattern painting of aircraft is of limited value when considering ground to air observation of low and medium-level aircraft.

a. Basic Color. The paint should be lusterless, and of the predominant color found in the natural terrain patterns in which the aircraft will be sited. In temperate zones and jungle areas and partially snow covered terrain, the color recommended is olive drab. In predominantly snow covered terrain, such as the arctic and antarctic areas, the craft should be completely white without a disruptive pattern. In desert terrains, because of the variations, no single color is recommended, but rather, desert sand, earth yellow, earth red, or mixtures of these colors.

b. Disruptive Patterns. Although the basic color provides a high degree of camouflage, in some cases, disruptive patterns add a marginal increase. The principles of pattern painting discussed in paragraph 24*c* are the same for aircraft—irregular, large, and bold, and of a color to blend with the terrain pattern. (Color and brightness contrasts not present in the background become conspicuous and should always be avoided.) Black is recommended for the disruptive pattern color in temperate zones, jungles, and deserts, except areas of the desert having light shadow, in which case olive drab is more effective. In partially snow covered terrain again olive drab is recommended. In the arctic and antarctic the craft should be completely white without any disruptive pattern. When technical considerations make it practicable, rotor blades on helicopters can be painted—the upper surface olive drab and under surface black. In arctic and antarctic areas the upper surface should be white. Figure 43 illustrates possible patterns. These are simply suggestions—no one set pattern should be followed religiously since variety is always desirable in any camouflage measure.

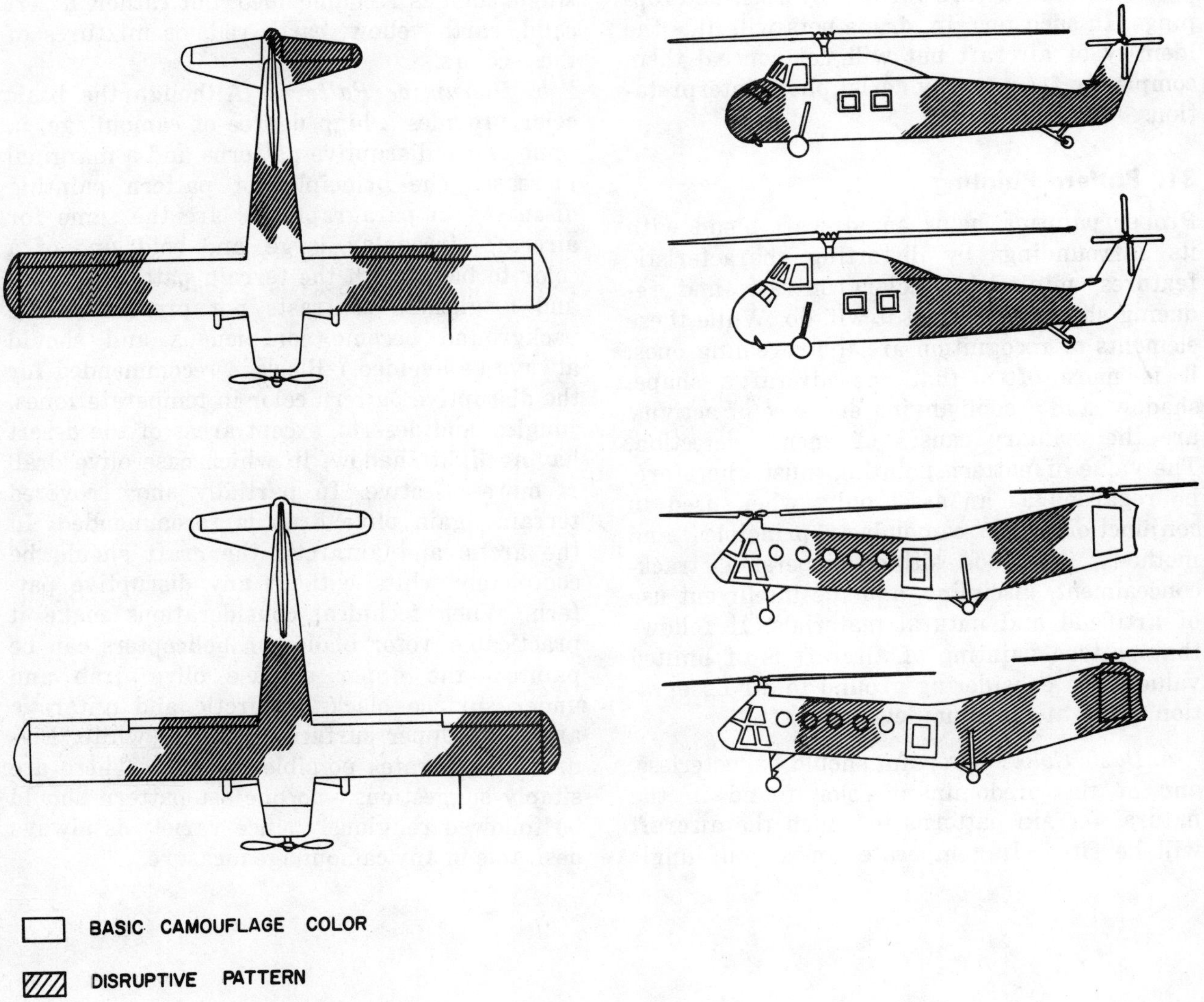

Figure 43. Suggested disruptive pattern painting for aircraft.

CHAPTER 8

BIVOUACS, COMMAND POSTS, AND SUPPLY POINTS

32. Introduction

A unit in bivouac is particularly vulnerable to enemy aerial observation and attack because its elements are concentrated in a smaller area than usual; and, except for the security elements, the men are resting and less alert than on the field of battle. It is at such a time that camouflage must be at its best (fig. 44). The unit must not only conceal itself quickly and efficiently, but must avoid the dangers of breaches in camouflage discipline.

33. Stages of Development of a Bivouac

There are four stages in the development of a

Figure 44. Aerial view of well-camouflaged bivouac area.

bivouac—planning, occupation, maintenance, and evacuation. Since it is often possible and probable that units must move without opportunity for planning, this stage may be lacking. In this case, the five points listed below in paragraph *a* must be satisfied in the area on and after arrival.

a. Planning. Frequent bivouacs are characteristic of modern mobile warfare. There is seldom time or facilities for elaborate construction; instead, bivouacs are quickly entered and quickly evacuated. No matter how swift the operation or how limited are time and facilities, the commander of a unit must plan for concealment in bivouac. The general area of the bivouac is determined by the tactical plan. Before going into this area, the quartering party should become familiar with the terrain pattern through a careful study of maps and aerial photographs, and be fully acquainted with the tactical plan and the camouflage requirements. There are five critical points for the party to keep in mind:

(1) Mission of the unit.

(2) Access routes.

(3) Existing concealment in the area.

(4) Size of the area.

(5) Concealment of all around defense element of the position.

b. Occupation. A carefully controlled traffic plan (para 23*a*(2)) must be rigidly adhered to while the units move into position. Guides posted at route junctions, fully aware of the camouflage plan, enforce camouflage discipline. Turn-ins must be marked to prevent widening of corners by vehicles. Foot troops must follow marked paths through the area. This is a critical period and bad camouflage discipline can negate any further effort at concealment. There must be no congestion of vehicles or activities and dispersion should be automatic. Seldom will vehicles be less than 30 meters apart in ordinary terrain or less than 100 meters in desert terrain. The three main congested areas—kitchen, maintenance, and the command post—must be dispersed.

c. Maintenance. Next to the occupation stage, the maintenance stage is the most critical. If the occupation has been successful from a camouflage standpoint, the maintenance is relatively simple. Successful maintenance involves frequent inspection of installations, active patrol measures for discipline, and when possible, aerial observation and photographs. Critical activities of a unit in bivouac are those that call for the congregation of troops, of which messing is the best example. It is here that the track plan must be rigidly enforced. Often it may be necessary to provide artificial overhead cover, such as flattops or drape nets. These, when used, must be carefully maintained. Garbage disposal pits must be concealed, with special care given to the spoil. Maintenance of night discipline is another difficulty. Men tend to relax at night. The same standard of camouflage discipline must be observed by night as by day, since night aerial photography will often reveal a unit that has become lax in this respect. Wired and taped paths must be followed. Blackout control must be enforced.

d. Evacuation. Camouflage measures taken at a bivouac do not end when the unit prepares to move out. An evacuated area can be left in such a state of disarray that aerial photos will reveal the strength and type of the unit, its equipment, and even its destination. It is an important part of camouflage to leave the area looking undisturbed.

34. Bivouacs in Barren Terrain

Experience on the desert has taught much about concealment in areas where large, convenient overhead cover is seldom found. Such areas, comparable to the desert as far as camouflage is concerned, are unplowed fields, rocky areas, grasslands, and other wide open spaces. The desert has taught that concealment in such areas is not impossible (fig. 45). Certain kinds of predominantly flat terrain have shadows cast by folds in the ground, large enough to allow some concealment by sitting alone. Judicious use of drape nets can render objects inconspicuous. Even in essentially barren terrain excellent concealment is possible when the configuration of the ground is irregular enough to produce a strong shadow pattern.

35. Bivouacs in Snow Covered Terrain

Although concealing a bivouac in snow covered

Figure 45. A bivouac area in North Africa during World War II.

terrain follows exactly the same principles as other camouflage, it presents several unique problems. A blanket of snow often eliminates much of the ground pattern, making blending difficult. Differences in texture and color disappear or become less marked. Snow covered terrain, however, is rarely completely white, and by taking advantage of dark features in the landscape, communication lines, streambeds, evergreen trees, bushes, shadows of snow drifts, folds in the ground, and the black shadows of hillsides, a unit on the move or in bivouac may often blend itself successfully into the terrain. Good route selection is usually more important than any other camouflage measures. Because of the exposed tracks, skis and snowshoes must not be used near the area since their marks are more sharply defined than foot tracks. To avoid tracking up an area, personnel, vehicles, and material should be restricted from the open areas. Bivouacs which have been well concealed in snow terrain for some length of time can be easily identified when the snow melts, unless precautions are taken. Compacted snow on much used paths melts more slowly than the uncompacted snow, leaving clearly visible white lines on a dark background. When this occurs, the snow must be broken up and spread out to quicken melting. The best way to minimize conspicuousness of tracks when moving or in bivouac is to follow communication lines or other lines which are a natural part of the terrain. Tracks coinciding with such lines are hard to identify. A turn-off from such lines must be concealed and the tracks themselves continued beyond the point. Windswept drift lines cast shadows and should be followed as much as possible. Straight tracks to an important installation must be avoided.

36. Command Posts

a. The command post is simply a specialized kind of bivouac, with a few additional problems of concealment peculiar to it. The command post is the nerve center of a military unit and because of this is a much sought target by the enemy. Command posts have functional requirements which result in the creation of characteristic signs, by which they may be readily identified. Some of these are:

(1) Converging communication lines—wire and road.

(2) Concentration of vehicles.

(3) Heavy traffic which causes widened turn-ins.

(4) New access routes to a position which could house a CP.

(5) Protective wire and other barriers surrounding the installation.

(6) Defensive weapons emplacements around the installation.

The camouflage solution to these problems is much the same as that for bivouacs. The primary factors are intelligent use of the terrain and background, and strict enforcement of camouflage discipline.

b. The site requirements of a large command post are essentially the same as for a good bivouac: preliminary reconnaissance and layout, quartering parties, rapid concealment of elements, camouflage discipline, and a well policed track plan to prevent visitors from violating it. There is one important additional consideration. A large headquarters is likely to remain in an area for a greater length of time than a bivouacked unit. It is for this reason that the site must be capable of being continuously occupied while offering a minimum chance of being disclosed by changes in the terrain pattern. It is unwise to locate a headquarters in the only large building within an extensive area of military operations. It is too obvious a place for such a post whether signs indicate it is being used as a headquarters or not, and is likely to draw enemy fire. If the command post is located in a building, there must be enough other buildings in the area to prevent pinpointing the target.

c. Communications are the life blood of a command post. Command posts sited to take advantage of existing roads and telephone and telegraph wires are easiest to conceal since new communications need not be created and the terrain can remain unchanged. When new communication means must be created, natural cover and terrain lines are used (figs. 46 through 51).

d. After the site has been selected and after camouflage has been erected to supplement whatever natural concealment is available at the site, continued concealment depends on discipline. Tracks as always must be controlled, vehicles should, if possible be parked several hundred meters from the command post. Security weapons and emplacements must be concealed; tracks to them must be inconspicuous. All spoil must be concealed. Protective wire and communication wire must follow terrain lines and be as well concealed as possible. Night blackout discipline must be rigidly enforced. Routes to parking areas for visitors must be maintained in accordance with the track plan.

e. In open terrain where natural concealment is afforded only by small scrub growth and rocks, overhead cover can be obtained by using drape nets or flattops. Even in desert terrain, broken ground and scrub vegetation form irregular patterns with which artificial materials may be blended. Digging in reduces shadow and silhouettes, and simplifies draping of emplacements or tents. In open terrain dispersion is particularly important. Routes between elements must be concealed or made by indirect courses—never in straight lines.

f. Headquarters in existing civilian structures presents the problem of hiding movement by day and concealing the evidence of activity at night, when blackout conditions usually prevail. Military movement in a village or a group of farm buildings is not easily discovered if kept to a minimum. Attempts to alter the appearance of buildings by disruptive painting is evidence of occupation and simply reveals a military installation. Erection of a small structure simulating a new garage or other auxiliary civilian building is unlikely to arouse suspicion, but any major changes will be closely scanned by enemy air observers. When buildings are partially destroyed and

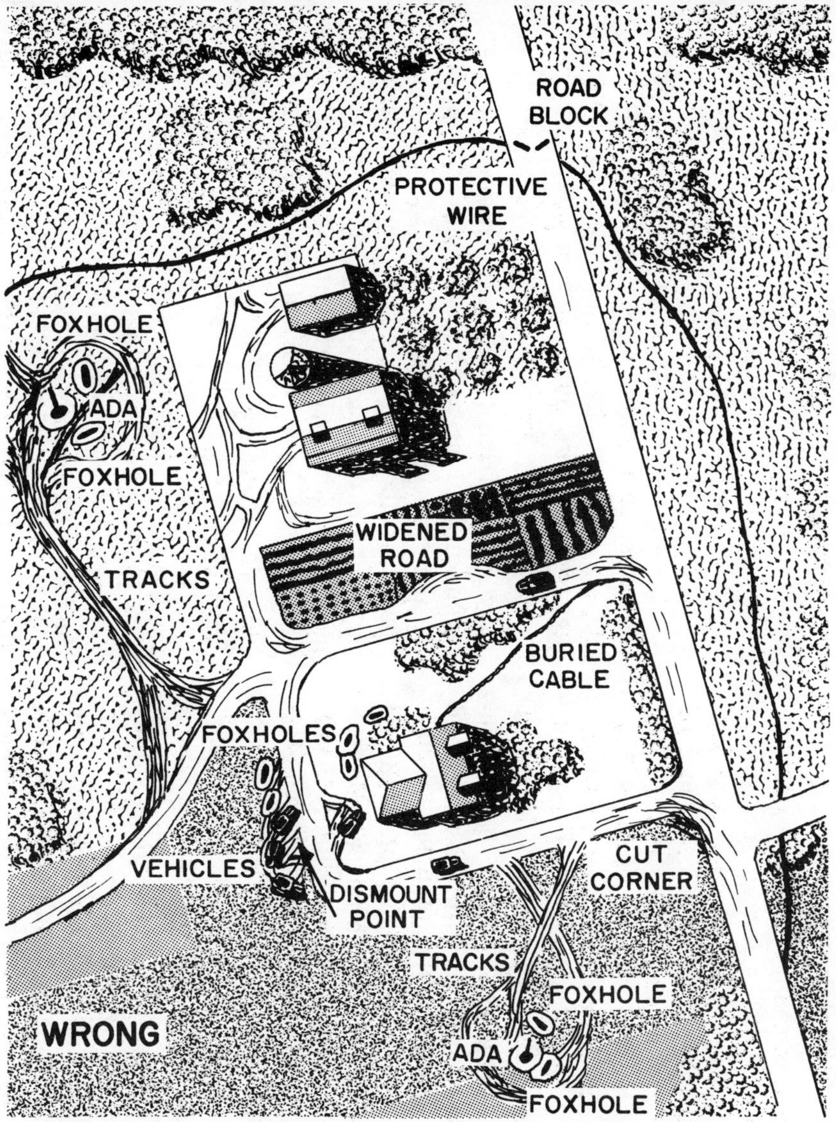

EVEN A HASTY GLANCE AT THIS SCENE WOULD TELL THE ENEMY OBSERVER THAT THIS IS A COMMAND POST. ALL THE TELLTALE SIGNS ARE THERE. THEY ARE MILITARY MARKS ON AN OTHERWISE ORDINARY RURAL SCENE.

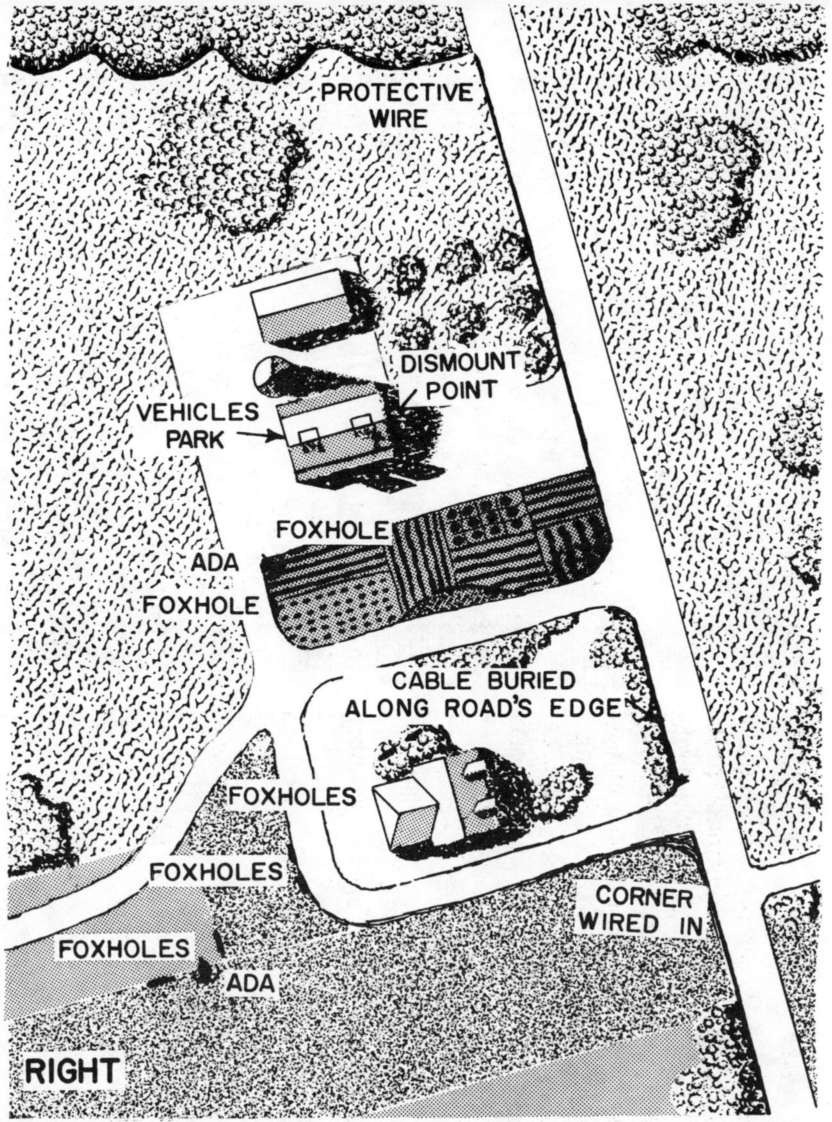

THERE IS NOTHING IN THIS RURAL SCENE TO AROUSE SUSPICION. SUCH A CONTROLLED CP IS POSSIBLE ONLY IF A CAMOUFLAGE PLAN HAS BEEN MADE IN ADVANCE OF OCCUPATION AND FOLLOWED CLOSELY. MOST VEHICLES SHOULD BE PARKED UNDER COVER AT A DISTANCE FROM THE CP. PERSONNEL SHOULD PROCEED ON FOOT TO THE BUILDING ITSELF. PROTECTIVE WIRE FOLLOWS TERRAIN LINES.

Figure 46. Layout of a command post.

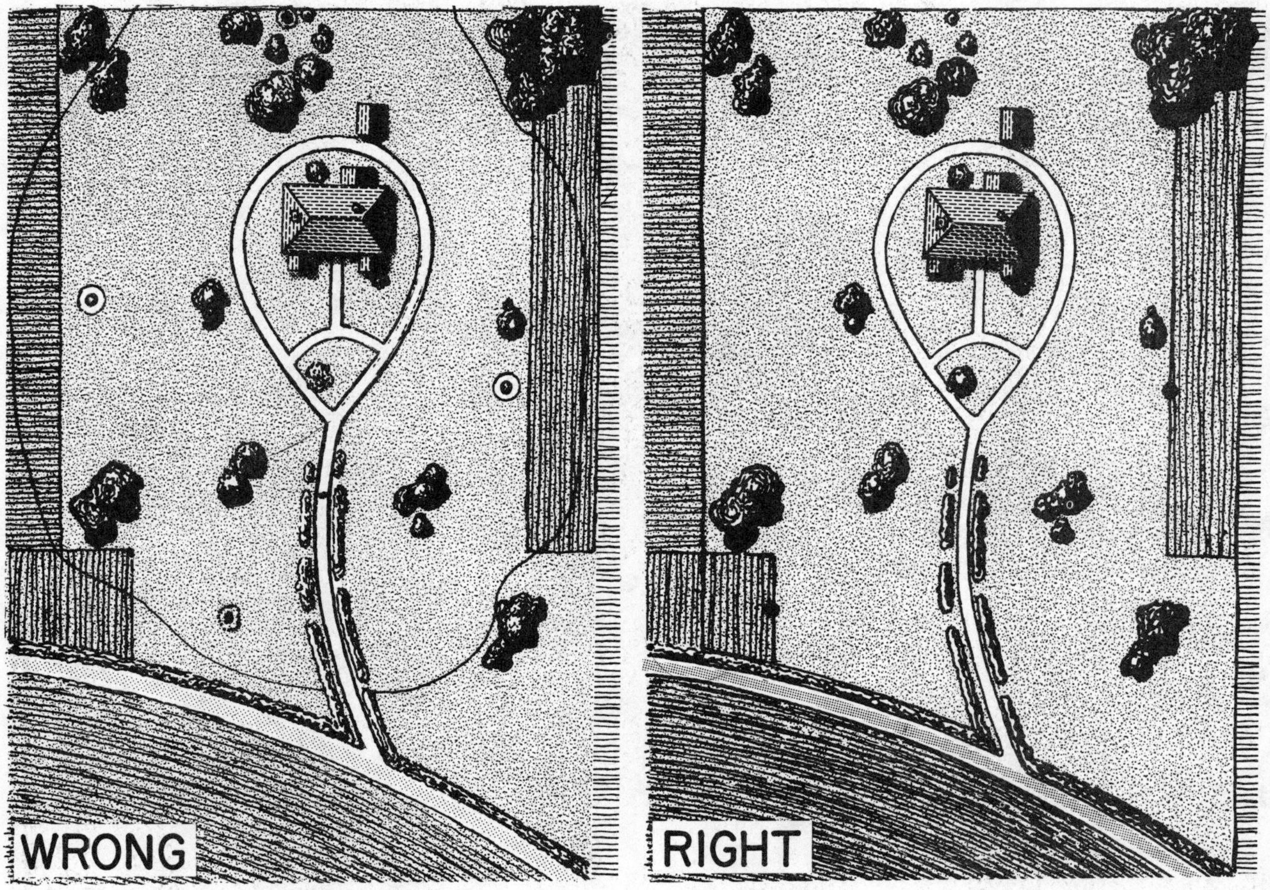

Figure 47. Proper layout of protective wire.

debris-littered, installations may be camouflaged with debris to blend with the rough and jagged lines of the surroundings. A few broken timbers, pieces of lath, plaster, and scattered rugs will accomplish quick and effective concealment. Other debris usually available includes rubble, scrap metal, wrecked vehicles, and furniture.

37. Supply Points

a. The problems in camouflaging supply points include all the difficulties of both bivouac and command post concealment, plus a number of particularly troublesome factors peculiar to supply points alone. Supply points vary in size from large concentrations of materials in rear areas to small piles of supplies in the forward areas. From a camouflage viewpoint, the large concentration of materials is the main problem. Huge amounts of equipment of all kinds are brought up quickly, must be unloaded and concealed quickly, and yet must be easily accessible for redistribution. Flattops are an effective solution if the supply points are not too large, if time and materials are available, and if they can be made to blend with the terrain. For supply points that cannot be concealed, decoy points will often divert the force of an enemy attack.

b. Supply points make use of natural cover

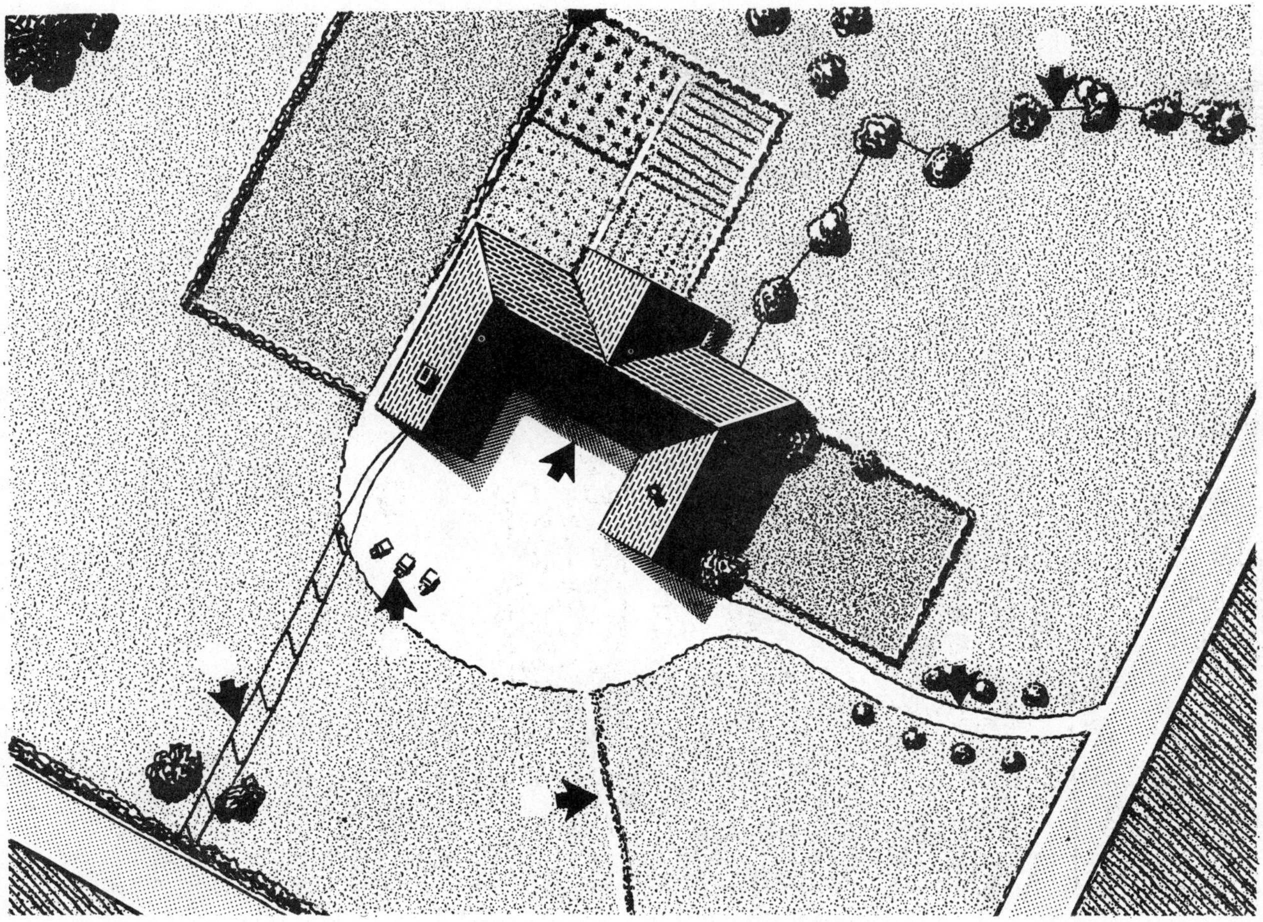

Communication wires are concealed by running along terrain features to destination. Scars made by burying cables are concealed by following road's edge. Wire may be strung from tree to tree, or, when poles are necessary, they should be placed along existing terrain lines. Maintenance crews especially must beware of making tracks which converge toward command post.

Figure 48. Proper layout of communications wire.

and concealment whenever possible. Stacks of supplies are dispersed to minimize damage from a single attack. New access roads are planned to use existing overhead cover. In more permanent installations, tracks running through short open areas can be concealed by overhead nets slung between trees. Traffic control includes measures to conceal activity and movement at, to, and from the installation. Even when natural cover is sparse or nonexistent, natural terrain features are used to advantage. There are few kinds of terrain which will not have to be used to store supplies and there are few kinds of terrain for which there is no solution. In cultivated fields, supplies can be laid out along cultivation lines and textured with strip-garnished twine nets to resemble standing stubble. Figure 52 illustrates how supplies can be stacked for deception. The plowed field usually spells trouble to attempts at camouflage, but supplies can be stacked parallel to the furrows and covered with earth colored burlap. From the air, this is effective concealment. Access routes are made along the furrows and no unnatural lines appear on the pattern.

c. Camouflage discipline measures at supply points include track plans that result in a minimum of changes in the appearance of terrain, control of debris so that it does not accumulate and attract enemy attention, con-

Figure 49. By continuing existing road (dotted lines) a traffic loop is created. This makes a conspicuous turn around unnecessary. It also provides access to vehicle park concealed in woods at a distance from the CP.

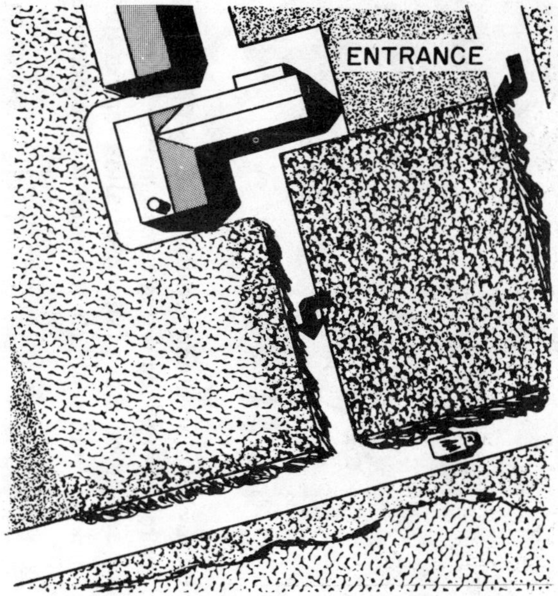

Figure 50. Small parking area for visitors. Overhead cover is created by placing natural materials in wire netting. Unless overhead cover is available, visitors must not be permitted to dismount near command post.

Figure 51. Tracks must be controlled and all traffic restricted to existing paths and roads. Edges of road turn-offs, intersections, and short narrow roads subject to heavy military traffic are wired in to prevent conspicuous road widening and corner cutting.

Figure 52. Stacked supplies.

cealment and control of trucks waiting to draw supplies, and maintenance of camouflage measures.

38. Camouflage of Water Points

a. Factors which aid in the concealment of water points are:

(1) Adequately concealed road net at point.

(2) Sufficient natural concealment to hide waiting vehicles.

(3) Adequate concealment—artificial or natural for operating personnel, storage tanks, and pumping and purification equipment.

(4) Strict enforcement of camouflage discipline.

(5) Control of spilled water; adequate drainage to prevent standing pools of water which reflect much light.

b. Foliage not sufficiently thick for perfect concealment is supplemented by natural materials, flattops, or drapes. Concealment is required for water point equipment; the shine of water in the tanks; and small open areas that must be crossed by vehicles or personnel in operating the point. Shine on water can be concealed by canvas covers or foliage and the characteristic shape of tanks can be distorted by foliage or artificial materials.

c. Camouflage discipline at a water point requires a water supply schedule for using units. Lack of a schedule, or violation of a schedule, usually produces a concentration of waiting vehicles which cannot be concealed.

CHAPTER 9

LARGE SCALE AND RELATIVELY PERMANENT INSTALLATIONS

39. Introduction

The principles and methods of camouflage stressed throughout this manual are applicable to the camouflage of fixed and relatively permanent installations. Any difference is merely one of degree of detailed planning and extent of camouflage construction projects. Given the time, material, and labor, there is almost no limit to the concealment that can be accomplished, if the importance of the installation justifies such expenditures. If its importance does not call for all-out camouflage, its visibility can still be reduced materially by the simple method of tonedown. This chapter presents suggestions for analyzing camouflage problems, preparing designs, and supervising projects. TM 5–200 covers the construction materials and camouflage techniques which can be modified to suit variations in specific installations.

40. Desired Quality of Camouflage

The kind and degree of camouflage desired is determined by the following factors:

a. Importance of installation (how difficult it would be to replace).

b. Vulnerability of installation (dispersion and susceptibility to damage).

c. Probable enemy knowledge and evaluation of installation.

d. Extent and efficiency of air-warning facilities and degree of air superiority.

e. Probable heights, directions of approach, and times of enemy observation or attack.

f. Probable angles of enemy observation.

g. Average visibility (rain, fog, and other atmospheric conditions).

41. Restrictive Factors

Factors which may limit camouflage possibilities are:

a. Prominent landmarks which serve as reference points for enemy pilots. Practicality of obscuring these landmarks should be considered.

b. Normal operational demands of installation.

c. Area involved and time allowed.

d. Expected useful life of installation. This influences decision on short or long range camouflage program.

e. Absence of suitable area for a decoy.

f. Availability and types of camouflage materials, labor, and equipment.

g. Seasonal changes and expected maintenance, determined from study of year round weather conditions (rainfall, temperature ranges, snowload, and wind).

h. Probable enemy use of aerial photographs of area.

i. Security requirements, including secrecy during construction.

j. Nature of adjacent installations, especially with respect to existing or planned camouflage. This is important when such installations are under control of another authority.

k. Degree of cooperation to be expected from units whose activities will affect success of final camouflage scheme.

42. Procedures

Before formulating a final plan for a project, a decision must be made as to the best construction procedure to follow. The following list suggests some of the possibilities. The final choice depends on the situation that will exist at the site at the time the work is initiated.

a. Complete camouflage applied step by step during construction or during a halt in the operational activities.

b. Complete camouflage applied to successive sections of the installation during operational activities.

c. Complete camouflage applied first only to vital parts of a large installation.

d. Hasty temporary camouflage measures to be replaced or augmented by more comprehensive and more permanent work.

e. Simultaneous development of completely camouflaged installation and decoy.

f. Construction of a day or night decoy while real installation is given tonedown treatment.

43. Essential Reference Data and Aids

The materials which should be gathered prior to the plan and which should be used as reference and aids in creating the plan are:

a. Medium and large-scale topographic maps of the immediate and adjacent areas.

b. Aeronautical charts.

c. A controlled mosaic, at a scale not smaller than 1:25,000 of the project area; and a controlled or semicontrolled mosaic at a scale not smaller than 1:50,000 of the surrounding area.

d. Aerial vertical photographs of the project area with a minimum overlap of 60 percent and oblique photographs, taken from cardinal directions or most likely approach angles.

e. Town plans and country maps.

44. Schedule of Operations

A schedule of operations should be prepared early in a camouflage construction project. This schedule should be planned so that:

a. The project does not interfere at any time with the functioning of the installation.

b. Materials can be ordered and deliveries scheduled to avoid a storage problem.

c. Different types of work do not conflict by being carried on at the same time in the same area.

d. One type of work can be substituted for another in the case of unforeseen delays.

45. Layout Grid Control

a. A plan of the area to be camouflaged should be drawn at a scale of not less than 16 feet to the inch, and should be gridded at 4-foot intervals, using the modular system. The 4-foot interval, or module, is a unit of measurement for regulating proportions. Its use as the basis for a plan will reduce building costs, offer the designer a simplified method of dimensioning drawings, eliminate the necessity for much expensive detailing, and offer a system of repetitive module, 4 feet in dimension, on which repetitive aspects of the camouflage can be constructed by an assembly line type of production. The 4-foot interval coordinates the sizing of different materials on a common basis so that when assembled they can be readily fitted together to form a complete structure. The better the components from different manufacturers can be fitted together, the less will be the cutting and adjustments required on the job. Planning for the use of modular products does not hamper designers in creating camouflage construction to meet any need. It simply means that designers, producers of building products, builders, and craftsmen all work together on a common basis using a coordinated system of dimensioning.

b. Preparation of working drawings on a modular basis (fig. 53) is not essentially different from that customarily followed in architectural practice. However, a new factor has been added—the discipline of the grid.

(1) *The modular grid.* Coordination of building products in a structure is based upon a 4-inch cube represented as a 4-inch grid on plans, elevations, and sectional drawings.

(2) *Small scale drawings.* At scales of less than 1 inch to the foot, it is not practical to show grid lines. An architects scale permits drawings to be laid out in multiples of 4 inches. Plans and elevations for camouflage construction are to be laid out using a 4 inch grid.

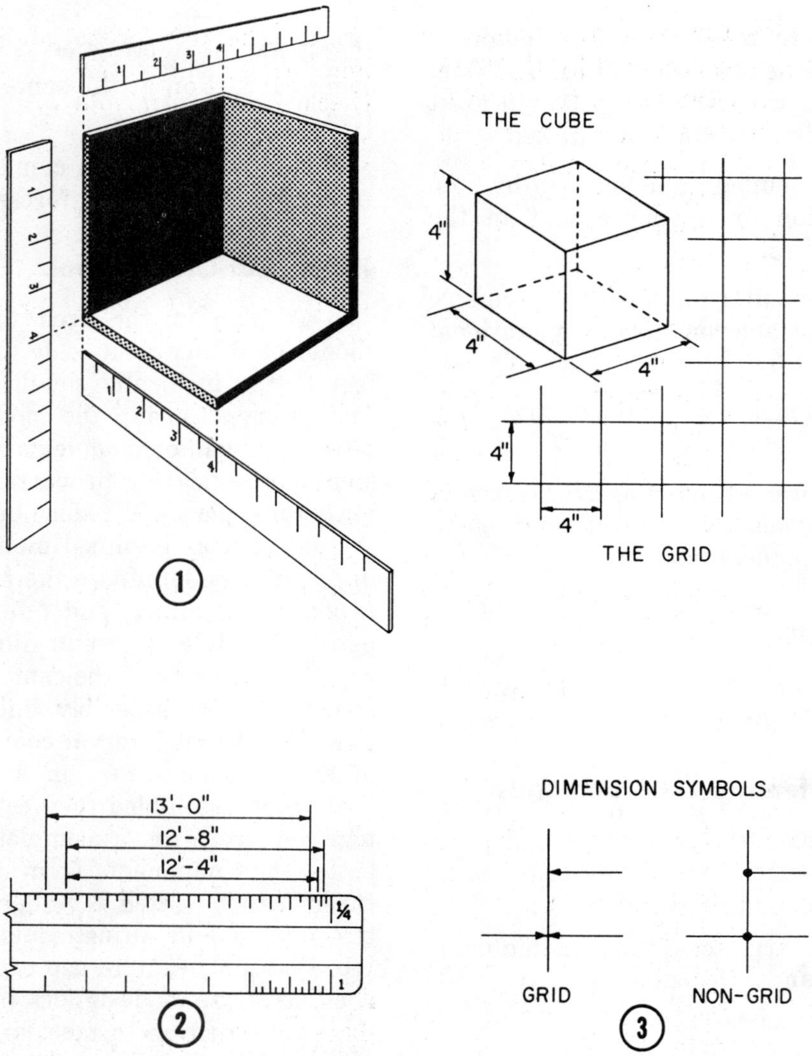

1 The 4-inch module and grid.
2 The architect's scale allows multiples of four.
3 Dimension symbols.

Figure 53. The 4-inch module as used in drafting.

(3) *Modular details.* A 4 inch grid is used in drawing typical details at a scale of 3 inches or 1½ inches equaling 1 foot. The grid is the basis of coordination and not necessarily a dimension of materials. Materials are shown as actual size and either located on, or related to, a grid line by a reference dimension. Dimensions on grid lines are shown by arrows; those not on grid lines by dots.

46. Marking the Area

After the layout plan is determined, a grid of 16-foot squares (4 modules) is transferred to the ground. Lines marked on the ground must not vary more than 1 foot from the design to maintain the scale of the design. Variations in scale make it difficult to match patterns at side walls and roofs and between areas. Lines can be marked on the ground using a tennis court marker, chalk lines, or any other marking device. With the grid lines as guides, the pattern outlines are then drawn.

47. Discipline

Camouflage discipline is vital during all construction phases at any site to be camouflaged. Effective discipline requires constant supervision when construction is designed for a large area. Clearings for buildings should be limited to the area to be occupied by the building. Building locations can be shifted slightly to avoid cutting down trees that will make subsequent camouflage easier. Scrap lumber, packing boxes, empty paint cans, and other forms of refuse and debris, as well as the spoil from excavations should be disposed of or camouflaged as soon as possible. Parking areas are best dispersed and concealed and waiting points and turnarounds should be marked. Equipment not in use and stockpiles of supplies must be concealed or removed from the site. Working equipment must be screened for security. To prevent scarring the earth around small concrete structures, such as pillboxes, raised platforms can be used for concrete mixing, supplies, and spoil. All personnel must be familiar with the plan as it concerns their own individual activities (figs. 54 and 55).

48. Inspection

During construction, materials should be constantly checked for suitability, quality, color, and proper application. The paints should be checked for color and type; cotton nets and wire netting should be inspected before use, with particular attention given to the garnish. The overall construction plan should be checked frequently on the ground and also by aerial photographs for indirect analysis. Frequent night inspections are valuable to discover any violation of the principles of good camouflage.

49. Camouflage of Buildings

The basic methods of concealment—blending, hiding, and deceiving—can be applied either to existing buildings or to new construction. Concealment is much easier, however, when the camouflage scheme is incorporated in the designs for new construction.

a. Disrupting Shape and Shadow. The shape and, to a limited degree, the shadow of buildings can be disrupted by pattern painting the

Figure 54. Careless and widespread earth scarring is caused by failure to establish a traffic plan, indescriminate use of earthmoving equipment, and too much concern with leveling and clearing to facilitate construction. An enforced traffic plan, set up during the planning stage would eliminate much of the revealing scars in this picture.

Figure 55. When a large area of woods must be cleared, as required for an airstrip, the trees should be cut in irregular patterns rather than in straight lines, even though this procedure might require more time and labor.

walls, the roof, and the surrounding ground. Large irregular patterns of two or three colors simulating the local terrain pattern can be applied in such a way as to break the straight edges. Because the roof reflects more light than any other part of the building, the pattern should be darker and the roof should be textured before being painted. The dark patterns on the roof are carried down onto the wall surface to break the line of the structure (figs. 56 and 57). The ground can be sprayed with black bituminous emulsion to break the shadow; coarse textured materials such as cinders, slag, or coal washer refuse can be spread matlike around a building in an irregular pattern to obscure the shadow; or thick shrubbery or trees can be planted, if practicable, close to the sides of a building. Rigid silhouettes added to the eaves of the buildings will distort their shape (fig. 58).

b. Digging. If the terrain permits, a new structure can be partially dug in, in order to reduce the height and in turn the shadow (fig. 59). Buildings of metal or those housing metal materials may be concealed from radar and infrared detectors by piling dirt in a gradual slope up to the eaves and placing approximately 3 inches of soil on the roof. (The moisture content of the soil used in this manner must be kept at approximately the same as that of the surrounding soils, or a reflectance difference will be created to negate the concealment value.)

c. Screening. Buildings can be concealed by screens of garnished nettings (figs. 60 through 62). Where concealment from close observation is required the netting should be sloped gradually to the ground. Disruptive patterns may be painted over netting, roof, and gable-end walls. For structures with roofs steeper than 30°, the netting must cover the buildings.

d. Disguising. The nature and size of buildings can be disguised in many ways (figs. 63 through 70).

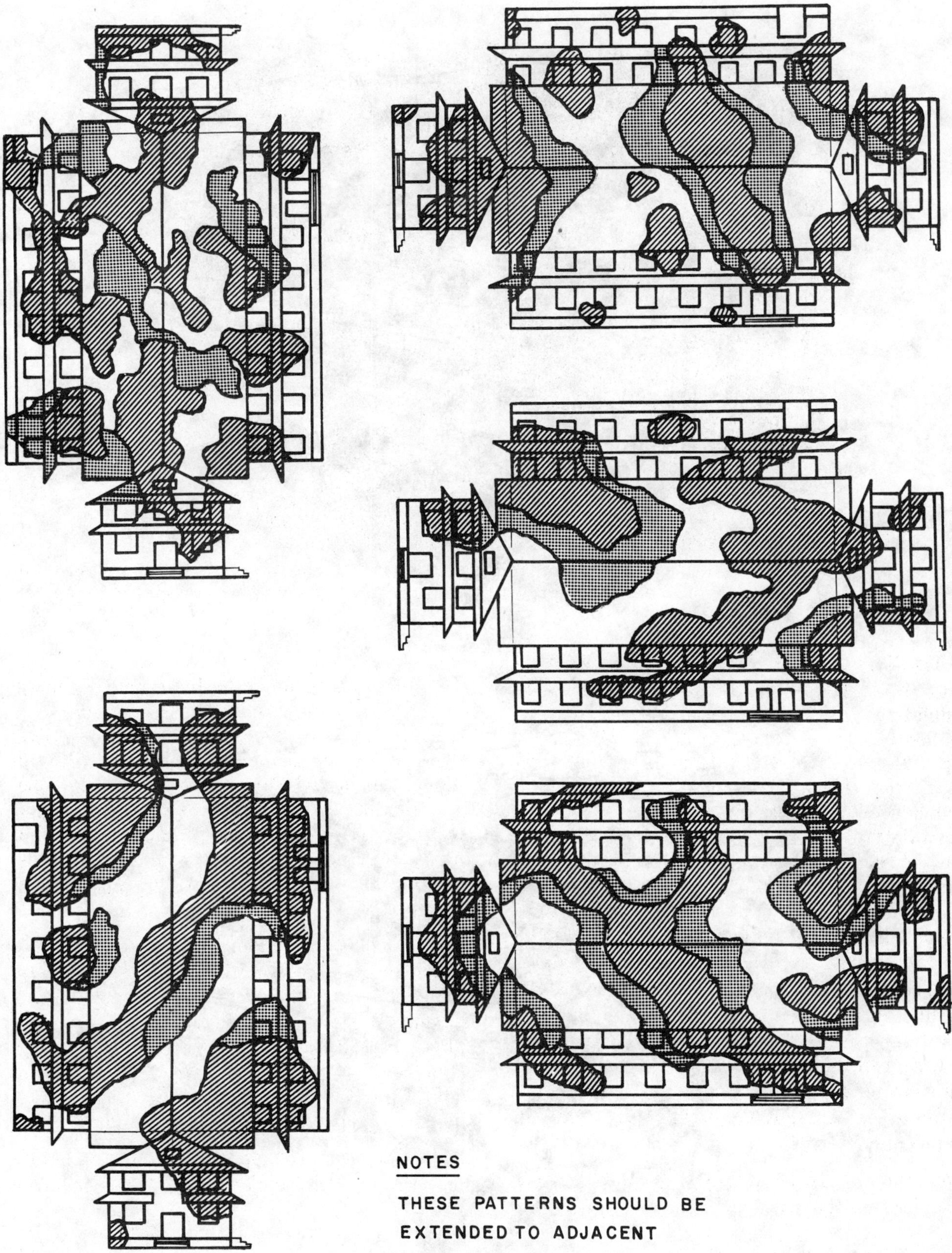

Figure 56. Examples of pattern painting for small buildings.

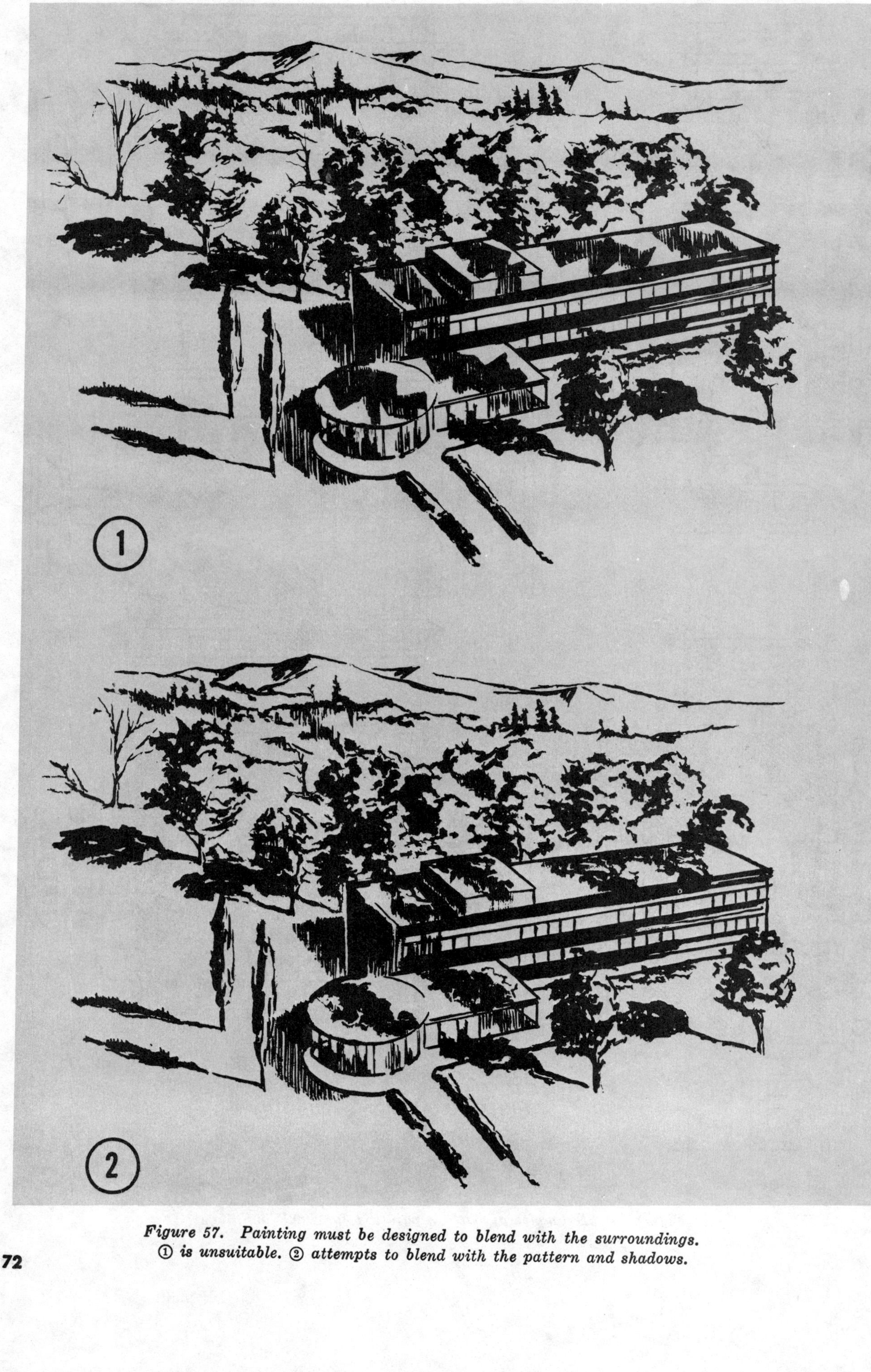

Figure 57. Painting must be designed to blend with the surroundings. ① is unsuitable. ② attempts to blend with the pattern and shadows.

Figure 58. Another method of breaking identifying shadows is to attach silhouettes of plywood or other rigid materials to the eaves.

Figure 59. Here is a quonset hut blended into the terrain by a combination of digging in, texturing, and covering with a net.

Figure 60. The jungle has been extended by a garnished netting which completely conceals an operations building from aerial observation.

Figure 61. Netting garnished with steel wool distorts the shape of this building from aerial view.

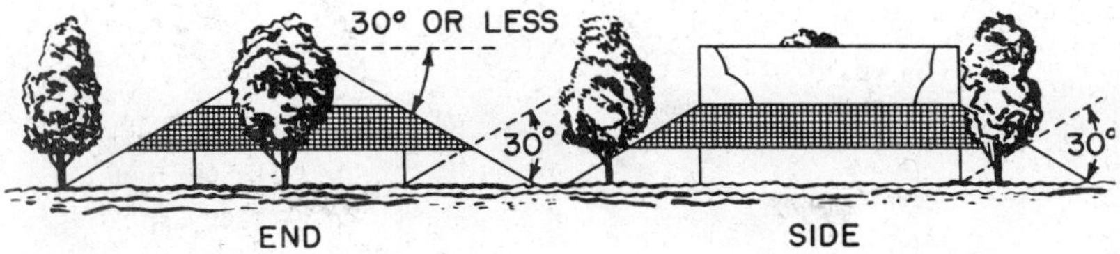

Figure 62. On small buildings where the slope of the roof is 30 degrees or less, the netting runs completely around the building, starting at the eaves and extending only far enough beyond the eaves to mask the ground line of the building when viewed from an angle 30 degrees above ground level.

Figure 63. Long military structures are made to look like several small buildings. Shadows and trees between the buildings are simulated by texturing and painting. The portions of the roofs left exposed are painted to match roofs of actual nearby houses. All the walks and paths lead directly to simulated entrances and have been textured to blend with the real ones.

Figure 64. Depot warehouses, power houses, and similar large structures in urban areas are often made to appear to be a group of smaller buildings characteristic of the locale. Roof lines are varied with wooden framework covered with burlap or fine mesh wire netting to simulate sloping hip roofs. If pavement prevents planting real shrubbery, two or three dimensional false shrubs of steel wool or glass fiber will disrupt the shadows. Firewalls projecting above a roof can be transformed into garden walls or hedgerows on simulated property lines. Monitors, elevator shafts, and other projections can be disguised as small buildings.

Figure 65. Structures can be made to conform to the architecture common to the locale by erecting superstructures of light timber and covering them with burlap, plaster, or other material. Here, a quonset hut is being altered to resemble the surrounding native houses.

False trees at various points around the tent conceal effectively from aerial view.

Figure 66. *Several methods, effective for disguises and concealment of tents.*

A frame covered with feather garnished wire to simulate a tree erected over the tent.

Figure 66—Continued.

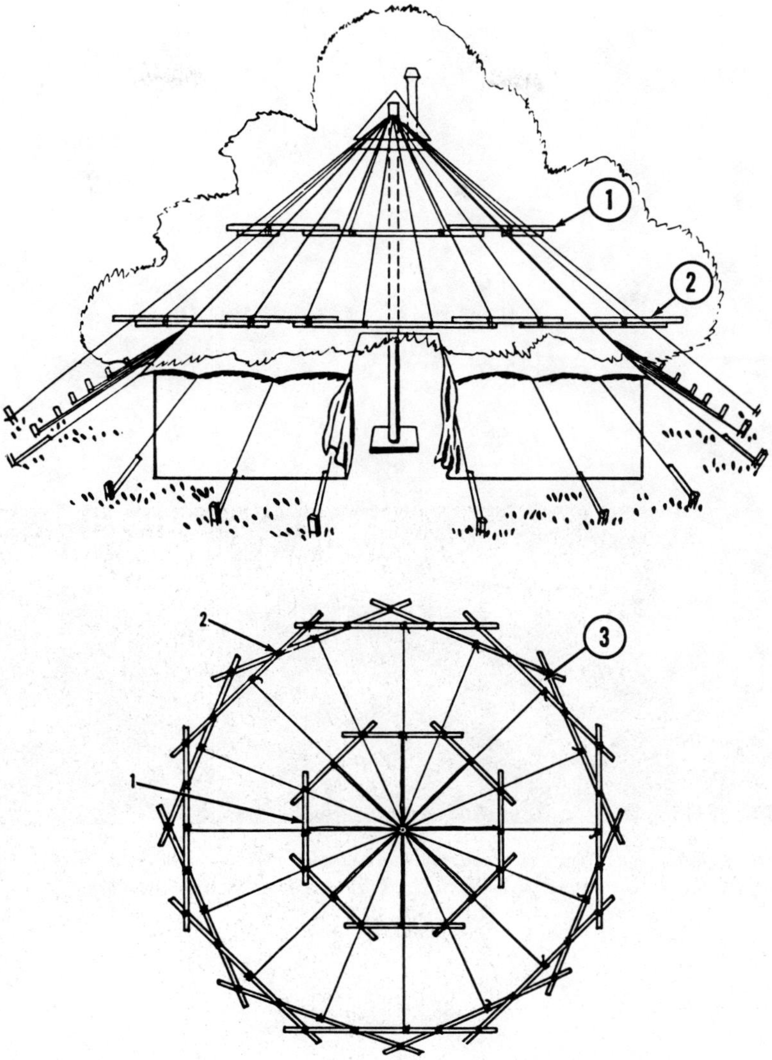

1 Frame upper.
2 Frame lower.
3 Location of false trees.

More elaborate construction of 1 x 2-inch lumber suspended from crown of tent by wire with false trees placed as indicated requires little maintenance.

1. Frame upper 2. Frame lower 3. Location of false trees

Figure 66—Continued.

Figure 67. *Disruptive pattern painting is used in conjunction with false structures. Houses similar to those in the background are simulated by three dimensional false roofs.*

Figure 68. Ground view of a camouflaged messhall ofa munitions factory in Brisbane, Australia (1943). The false road continues on over the building and terminates logically at a route juncture. This is an excellent example of camouflaging a permanent installation.

Figure 69. The center area ① shows the typical symetrical barracks. On both sides ② the proper dispersion of barracks, in wooded areas, combined with tonedown, texturing of roofs, and pattern painting is shown.

Figure 70. Closeup view of barracks shown in ②, figure 69.

50. Roads and Parking Areas

Roads can be completely concealed for comparatively short stretches only. However, intersections, traffic circles, short access roads, and parking areas, when they are landmarks, can be made inconspicuous by tonedown, texturing, screening, transplanting trees, relocating roads, or by making decoy roads. Tonedown painting reduces the distance from which a road can be seen. How effective this method is depends on how closely the texture and color used match those of adjacent areas. It is difficult to obtain the proper tone value by painting alone, texturing the surface will help (fig. 71).

51. Railheads, Railways, and Rolling Stock

All of the revealing characteristics of the activities and construction associated with a railroad are extremely difficult to camouflage. The railways themselves are recognized by their long parallel lines, gradual curves, light colored ballast in the roadbeds, and the shine from mainline tracks. The railheads are recognized by their size and the attendant activity. Camouflage measures, at best, can make the installation appear unprofitable to the enemy. When there is a choice, the site should provide as much natural overhead cover as possible for spur tracks, truck parks, storage areas, warehouses, and access routes. The railhead itself should be at some intermediate point, not at the end of the rail line. (False tracks leading to a decoy railhead is probably the best solution for confusing the enemy observation and attack.) The junction with the mainline should be as inconspicuous as possible. Specific techniques are—

a. Adding extra ballast to cover ties.

b. Making outer edges of ballast irregular.

c. Placing fitted screens between and on sides of rails.

d. Erecting netting over sidings, between building to conceal loading platforms, or over access routes to storage areas.

e. Enforcing a track plan for vehicles will keep visible signs of unrelated activities at a minimum.

f. Camouflage of supplies stored in the open.

g. Dispersing freight cars, locomotives, vehicles, and supplies. This is a normal precaution and is essential when other camouflage is impossible.

1. Exterior view 2. Interior view

Figure 71. Parking area screened from observation.

52. Bridges

While a bridge is difficult to hide, there are a number of camouflage tricks which may deceive the enemy as to its location and, most important, its condition. Simulated craters can be painted on the decking and covered. After an attack, they can be exposed. Decoy tanks and other decoy vehicles can be placed to simulate a traffice tie-up after the attack. Portions of the side railing can be removed. For this type of deception to be convincing, it must appear that an alternate river crossing site is being prepared. Approaches must be well worn on both banks. If there is a line of near-miss craters near the bridge, one or more of them should be filled. If a bridge has been actually damaged, it may be made to appear repaired and usable by filling a gap with wire framework covered with cloth. Then it again becomes a logical crossing and protects the actual crossing. If a riverbed is suitable and the water slow and muddy, bridges may be constructed with the deck submerged just below the surface. The shape and shadow of a destroyed bridge may be used to help conceal a ponton bridge constructed alongside. If a decoy crossing is built at a logical place, some distance away, the expedient has a greater chance of escaping detection. In certain cases it may be possible to submerge a ponton bridge which has pneumatic floats by extracting air from the floats. By this means, the bridge can be hidden during daylight hours and refloated for use at night.

53. Wire Lines

A well camouflaged structure is of little value if a conspicuous line of communication wire terminates at the installation. It is disastrous to allow unconcealed cable lines to end abruptly at what is meant to appear as an innocent hill but is actually an important command post. A decoy must continue past the installation to a logical termination or the real line must be camouflaged. Imitation lines can be made of rope, wire, cord, or other similar materials. The presence of a line can be concealed to a great extent by carefully locating it along terrain lines. Irregularly sized supporting poles with the bark left on, set at irregular intervals and staggered to conform to the ground pattern, are less conspicuous than lines regularly spaced and alined. Spoil taken from the pole holes must be carried away or hidden. Care must be taken during maintenance to avoid making an obvious path along the line of poles.

54. Pipelines

Pipelines should be laid along secondary roads whenever possible. When cross-country laying is necessary, terrain features should be fully utilized. To eliminate the shadow of the pipe, dirt or debris blended with the background should be banked gently along both sides of the pipe. A tonedown color applied to the pipe helps blend it with the background. Tanks and pumping equipment should be recessed in pits, dispersed, and concealed by natural cover or nets. False pipelines are easily simulated by the use of ditching equipment; after each day's work, several poles should be left at the end of the ditch to simulate a stack of unlaid pipe.

55. Airfields

a. The first step in the camouflage of an airfield or landing site, as in all camouflage operations, is a thorough study of the terrain by air observation and aerial photographs. After the ground formation, predominant colors, and patterns of the area are analyzed, the camouflage scheme can be planned.

b. Dispersed parking sites should be located within dark and heavily textured parts of the area, avoiding sites where the aircraft would be in sharp contrast to their surroundings or close to an unusual or isolated terrain feature. Use should be made of overhead concealment, clumps of bushes, scrub growth, folds in the ground, and other shadow-casting irregularities.

c. Circulating traffic should be confined, if possible, to existing roads and paths for movement within the area and for access to security outposts. New routes should be kept to a minimum. Advantage should be taken of overhead concealment and vehicles should be sited close to and along natural terrain lines.

d. The greater the traffic at an airfield the more difficult it becomes to camouflage runways and taxiways. Sod airfields do not ordinarily present too great a problem unless heavy use has created worn paths and strips. If this does happen it may be possible to extend the path or strip into an already existing road or trail, thereby concealing its actual purpose. In this respect, cooperation on the part of the pilots by restricting their landings, takeoffs, and turnarounds to designated areas is vitally necessary. Hard runways and taxiways may be textured and painted to blend with the background and roads can be simulated across them.

e. The camouflage of revetments is difficult because of their contained shadow. If they are partially dug into the sides of hills, the problem is simplified. The more irregular the shape and the more gradual the slopes, the easier they are to camouflage. Earthwork revetments may be seeded to give them texture and color and to conceal new spoil. Further improvement may be made by planting small shrubs or vines to break up the form and shadow. If time and facilities permit, tree planting is an effective measure.

f. The operational structures associated with airfields are treated as any other building or position and are covered in other parts of the manual.

g. Panels and other landing aids are displayed only when required. When pilots are familiar with the landing area the aids are removed and displayed only for visiting aircraft. Field lighting must be concealed to prevent ground or oblique aerial observation. Light emitted from airfield lighting fixtures can be controlled by use of the combat hood attached to all fixtures for this purpose. Fixtures may be blended into the terrain pattern by use of natural materials or by painting and texturing to match the terrain color.

CHAPTER 10

DECOY INSTALLATIONS

56. Locations

Decoys must be located in logical positions, far enough away from actual targets to prevent enemy fire on the decoy from hitting the real installations. In all cases the location of real installations must be carefully considered. The distance depends on the size of the installation, the type of enemy observation, and fire expected. Decoy bridges, depots, railheads, and airfields may be 3 to 8 kilometers from the real object. Decoy artillery batteries may be less than 1,500 meters from the actual position or separated from the position by much greater distances depending upon the type battery being simulated, the type warfare, and the mission of the unit. To deceive the enemy, a decoy simulating a large rear area installation should have approximately the same relationship to nearby landmarks as the target itself, since landmarks will be used as enemy reference points.

57. Theory of Decoy Deception

a. Direct and indirect aerial study must be made of each installation that is to be simulated. There is no standard to follow as every installation has its own peculiar signatures; even two of the same type will have individual and unique characteristics.

b. A decoy installation must be so constructed that its disclosure appears to be the result of poor camouflage. There are various methods of doing this, such as exposed parts of a decoy, exposed tracks, incomplete concealment of shadows of decoys, or the improper use of surface texture and color. If a decoy draws attention from a real installation for but a moment, it serves its purpose. A decoy position which has been discovered to be such by the enemy may later be occupied as an actual position.

c. To be effective, the decoy installation must include features normally associated with the real installation and must be properly maintained. For instance, decoy planes on an airfield must be moved from time to time; in the case of decoy trucks, a few real trucks should be used to make tracks; in every case, indication of normal activity should appear.

d. Decoys intended to divert attention from real objects or installations are effective only when the real objects are completely camouflaged.

58. Signatures

a. The characteristic telltale signs of military activities are called signatures. Tracks are the most important and obvious signature of any military activity, with their distinctive features easily differentiated from similar civilian activities (fig. 72).

b. The various types of tracks may be simulated in the following ways:

(1) *Foot tracks.* The desired tracks should be made by actual foot traffic. Straw or hay may be scattered to give the effect of more extensive use. Tracks in a presumably occupied position must be constantly increased in wear and width.

(2) *Wheeled vehicle tracks.* The best method of providing wheel tracks is to run several vehicles through the area to create the illusion desired. Chains or logs may be dragged to create a greater scarring of the ground.

(3) *Tracked vehicle tracks.* It is desirable to use an actual tracked vehicle to make these

	TYPE OF TRACK		
	FOOT	WHEELED	CATERPILLAR
GENERAL APPEARANCE	NARROW WITH MANY FILAMENTS AND UNTRODDEN PATCHES. i. e. 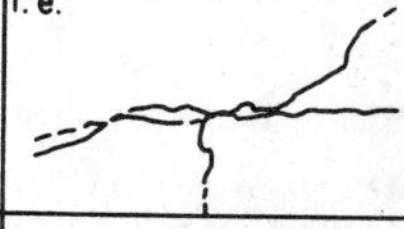	BROAD TRACK WHICH GRADUALLY COVERS AN AREA WITH A UNIFORM TONE.  EACH TRACK A SINGLE LINE	A NUMBER OF INDEPENDENT DOUBLE TRACKS CROSSING AND RECROSSING. i. e. 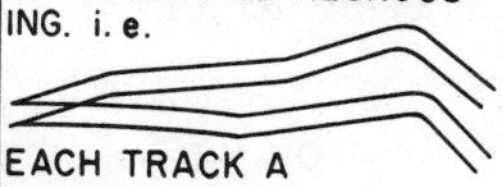 EACH TRACK A SINGLE LINE
STRAIGHTNESS	NEVER STRAIGHT.	STRAIGHT FOR SHORT DISTANCES.	WAVY BUT ALSO RUNS IN STRAIGHT STRETCHES.
BENDS AND TURNS	GOES AROUND SHARP BENDS AND ANGLES; NO SPECIAL MARK AT BENDS.	USUALLY AT LEAST A 20 FOOT RADIUS ON CURVES. CURVES EVEN.	MOST TURNS ANGULAR, CAN TURN AS SHARP AS 90°. U-TURNS CHARACTERIZED BY DEEP RUTS AT CURVE AND NOT AS EVEN AS WHEELED VEHICLE U-TURN.
HILLS	GOES STRAIGHT UP HILLS.	ZIG-ZAGS UP ALL BUT SLIGHT INCLINES.	MAY WIND UP A STEEP HILL - OTHERWISE GOES STRAIGHT UP.
CROSS COUNTRY	AVOIDS MINOR HUMPS AND DAMP SPOTS BUT GENERALLY NOT AFFECTED BY TERRAIN.	MUST HAVE GOOD GOING WITH EASY GRADES.	CANNOT GO OVER ROCK AND USUALLY FOLLOWS EASY GRADES IN EVEN TERRAIN.
NOTE: ALL TRACKS ARE IRREGULAR IN OUTLINE AND ARE MOST VISIBLE WHERE MANY TRACKS CONVERGE, SUCH AS AT GAPS IN WIRE OR MINE FIELDS OR TURN-OFFS INTO BIVOUAC AREAS.			

Figure 72. Track characteristics.

tracks, since they are impossible to duplicate accurately by any other means. Such tracks do not need to be renewed as frequently as the other two types.

c. Spoil is usually conspicuous near all dug positions. If, however, the quality of the camouflage discipline of surrounding troops is good and they dispose of their spoil, this same practice must be followed with the simulated units. On the other hand, if the camouflage discipline is not good and spoil appears around the dug positions, it must be reproduced at the decoy position. The best way to produce the appearance of spoil is to spread the earth from a real excavation, although the excavation need not be as deep nor the spoil piled as high as in the genuine position. The trampling of the ground by working parties flattens vegetation and compacts the ground so that the general effect on all but bare rock or sand is to make the ground surrounding the emplacement appear lighter in tone than its surroundings.

d. Shelters, such as dugouts or holes, show as dark spots in a light area of tracks and trampling. Spoil is also present. Airing blankets and similar items may also be visible. Tenting or shacks are easily improvised.

e. Latrines are an associated feature of every occupied site. They are usually disclosed by tracks converging and becoming more marked as time passes.

f. Buried cable is frequently an adjunct of important headquarters and may also be associated with radar installations. It appears as a track, usually straight with angular turns, and light in tone. With the passage of time its visibility decreases slightly.

g. Barbed wire is a feature of almost all infantry combat positions. The wire itself cannot be seen on aerial photographs, but its presence may be revealed by the tracks and trampling of the wiring party. After several days the location of the wire is disclosed by a faint gray line under the wire which gets darker and more pronounced as time passes because of the accumulation of untrampled vegetation. Gaps in the wire are often the means of disclosing its presence because of tracks which converge and diverge without apparent topographic reason.

h. Minefields, like wire may also be located on aerial photography by an otherwise unexplainable convergence of tracks, but they will more commonly be identified by the regular patterns of excavations which show up as light patches. Even minefields which are not laid to a standard pattern tend to show up on an aerial photograph, not because one mine hole is visible in itself but because a number of light patches concentrated in one spot draws the eye to that area.

59. Common Defects

Following is a list of the defects that most frequently cause a deception to fail. It will be noted that they are of general application and that any one of them may render worthless the otherwise most perfect decoy.

a. Regularity of tracks.

b. Lack of litter associated with military occupation.

c. Flatness (no stereoscopic relief).

d. Failure to faithfully simulate a particular type of installation.

e. Absence of motor transportation and lack of movement.

f. No daily change in appearance.

g. Incorrect tactical positioning.

h. Unreasonable speed of buildup or removal.

i. Lack of real air defenses.

j. Failure to simulate a necessary component of a particular installation.

60. Decoy Field Fortifications

a. Emplacements and Intrenchments. In simulating any dug-in type personnel or weapons emplacements, such as foxholes, trenches, and mortar or machinegun emplacements, the most satisfactory simulations are made by actually digging into the ground. However, it is not necessary to dig the decoys as deep as their real counterparts. A depth of at least 1 foot is usually sufficient to provide enough spoil for a parapet around the emplacement and to create the proper illusion of depth to the airman and aerial photo interpreter. A stronger illusion of depth may be obtained by partially filling the shallow emplacement with straw, brush, hay, or leaves. This breaks up the reflection of light from the bottom of the emplacement and appears from the air to be much deeper than it actually is (figs. 73 and 74).

b. Pillboxes. Pillboxes are located at the most likely avenues of enemy approach. The enemy knows this and can accurately guess their general location. However, if a pillbox is well camouflaged, blended into its surroundings, or disguised as a rock, bush, house, or other object which will make it inconspicuous in its particular setting, the enemy is prevented from spotting it accurately. To further direct the enemy from the real pillbox, decoy pillboxes may be constructed and insufficiently camouflaged so that the enemy can spot them with little difficulty (fig. 75).

c. Antimechanized Obstacles. Antimechanized obstacles are often located on reverse slopes, around curves, and in or behind natural screens to conceal them from the enemy and to gain surprise. Decoy antimechanized ditches are often effective in luring enemy tanks into real camouflaged ditches. Such a ditch need be only about 2 feet deep to create the proper illusion of depth and is constructed in the same manner as are trenches. Figure 76 illustrates a deception scheme using real and decoy antimechanized ditches and pillboxes together. Decoy dragon's teeth are also effective in guiding the enemy into real traps. They may be used in conjunction with real and decoy antimechanized ditches. Obstacles like dragon's teeth should not be simulated in areas not covered by antimechanized fire (fig. 77).

d. Minefields. A simulated minefield may be as effective an obstacle as a real field because the enemy must check each simulation to be certain it is not a real one. Such a minefield is effective against aerial observation. Simple ways to create a decoy minefield are: digging up the ground in a standard minefield pattern; erecting a minefield marking fence; and creat-

Figure 73. A decoy trench as it is being dug and as it appears close-up finished and filled with brush. Notice how the brush filling makes the completed trench appear much darker than the unfilled trench.

Figure 74. Another method to simulate an entrenchment is to place burlap or cloth in the outline desired and then fill in the simulated excavated portion with waste oil, paint, or scrapes of dark material. Note that the simulated parapet is formed by rolling under the edges of a large piece of burlap and creating the illusion of the mound of a parapet by piling sand under the burlap edges. The excavated portion of the slit trench is simulated in this figure by waste oil and dark colored salvage scraps.

Figure 75. Here is a decoy pillbox constructed of wood, covered with burlap or osnaburg, and sprayed with paint to resemble concrete. Sand and dirt may be sprinkled on the wet paint to give it a concrete-like texture.

ing the characteristic zigzag patterns of gaps through a minefield.

61. Artillery

a. Decoy artillery positions, in accordance with the basic principle of all military simulation, must conform to the positioning requirements of their real counterparts. Sometimes the camouflage of these decoy positions is difficult. In open areas the use of suitably garnished nets is highly desirable. If nets are not available, camouflage is obtained by properly positioning each piece of equipment to blend with surrounding ground features. For example, in orchards, each piece of equipment may be placed where a tree has been removed; in hedgerow country equipment may be placed in cleared spots in the hedge. When protecting an industrial area or a large installation which cannot be concealed, often no attempt is made to conceal the position, the theory being that the enemy will be cautious in attacking a heavily defended target.

b. Placing the dummy weapons and supplementary equipment in position is only part of the job of erecting a decoy artillery position. Tracks must be made around the position as in a real battery emplacement, because it is the tracks that most often disclose the position to the aerial observer. Without tracks, a weapon position looks fake; moreover, it would rarely be spotted if the weapon were well camouflaged.

c. Decoy positions may be simulated by digging in a shallow pit of the correct diameter and shape for a particular artillery piece and piling the spoil around it as a parapet or by forming a parapet of hay, straw, empty crates, old barrels, or loose dirt on the ground around the decoy. Even without the use of simulated guns or equipment, a decoy artillery position can be simulated perfectly merely by making blast marks and scattering debris. In snow terrain, blast marks appear black, and in clear terrain they appear light to dark gray. These are very convincing. One of the best and simplest ways of simulating an artillery position is to place partially camouflaged simulated weapons in vacated positions formerly occupied by real weapons. Flash simulators may be included with each simulation and used with the same regularity as the real weapons.

62. Bivouacs

a. When simulating bivouacs it must be remembered that each arm or service creates a distinctive appearance which must be reproduced to make the decoy convincing to the enemy air observer. The simulation of bivouacs is of particular importance in rear areas where reserves and fast moving units are generally located. Enemy air reconnaissance is partic-

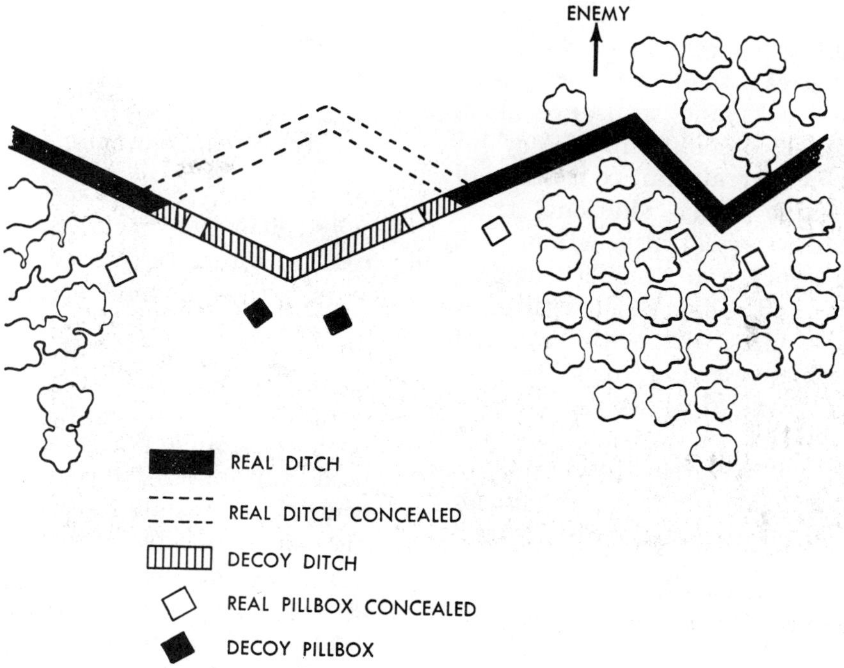

Figure 76. Layout of decoy and real antimechanized ditches and pillboxes.

Figure 77. Here is a row of decoy dragon's teeth and a decoy antimechanized ditch. A camouflaged real ditch is on the foreground, covered with wire netting, cloth, and sand.

ularly careful in scrutinizing areas behind the front for indications of buildup of troops or equipment which would indicate future attacks. Real bivouacs are generally located in areas that provide the best concealment for personnel and equipment from aerial and ground observation; dispersion; communications facilities; protection from attack in the vicinity of natural obstacles; and water supply.

b. From the air, the characteristics which most readily identify a bivouac are tracks, vehicles (types and sizes of which will identify the unit), paths, trails, and trash and litter caused by poor camouflage discipline. Bivouacs are more difficult to conceal in snow-covered, barren or desert areas, than in other types of terrain, because tracks and trails are almost impossible to conceal or obliterate. Decoy bivouacs in this type of terrain can be realistically simulated by making many additional tracks which do not lead to a real installation, or by digging shallow trenches and emplacements in the snow or sand and filling them with grass, leaves, etc. Brush piles with paths radiating from them resemble command posts or supply or ammunition storage areas. A good way to make a decoy bivouac is to examine aerial photos of a representative real installation and duplicate the track plan and some of the camouflaged objects in a simulated fashion but in a somewhat more conspicuous manner.

63. Command Posts

a. Some of the signs which indicate to the observer the possible presence of a command post are—

(1) Converging wire lines and vehicle tracks. Also, there may be various types of antenna arrays for radio communications.

(2) Concentration of vehicles.

(3) Heavy traffic causing widened turnins.

(4) New vehicle tracks to a position which could house a command post.

(5) Protective wire, foxholes, and other barriers surrounding the installation.

(6) Defensive weapons emplacements around the installation.

b. When it is impossible to conceal from the air the fact that a command post is in a certain area, then a decoy command post may be constructed in the vicinity. In this instance it is obvious that the decoy must look more like the real thing than its genuine counterpart, in order to make construction of any decoy worthwhile. Disguise of the genuine to look like a decoy may be feasible. Certain characteristic signs of occupancy should be made at the decoy including cross-country tracks simulating those made by a wire-laying detail, antenna arrays to simulate radio communications facilities, smoke and occasional lights, a few poorly camouflaged tents, new tracks from day to day, and vehicles in the area as well as other signs of activity.

64. Troop Concentrations

Various troop concentration points are generally identified by a collection of bivouac areas, motor pools, supply points, or entrenchments, with other associated activities.

65. Motor Parks

Around a motor or heavy equipment park there are innumerable scars, tracks, slit trenches, and a continual shifting of equipment. In simulating these installations, all such characteristics must be faithfully reproduced. Other items that may be required for deceptive displays are refuse piles, mess tents, bivouacs, latrines, and command posts. In desert or other barren terrain, motor parks or vehicle concentrations are best protected from enemy attack by dispersion over a wide area. Here the value of decoys is most apparent because real vehicles and equipment are extremely difficult to conceal or camouflage. Decoy vehicle concentrations are exceedingly valuable in drawing the enemy's attack and dispersing his effort.

66. Tank Concentrations

Decoys of armored parks and bivouacs should follow the same general techniques as described in paragraphs 62 and 65. Distinctive tracks made by tanks and other tracked vehicles both locate and identify the unit.

67. Airfields

There are two distinct types of decoy airfields, day fields and night fields. The day decoy field consists of prepared runways, buildings, access roads, huts storage, and the auxiliary features. The night decoy field consists entirely of lights, and should not be visible during the day. Normally the day and night decoy fields are separate installations.

a. Offset Distance. The distance between the decoy airfield and its real counterpart should be from 6½ to 12½ kilometers, but again depends on the type of warfare. However, a decoy situated more more than 12 kilometers from the real airfield is likely to be regarded as an entirely separate airfield and cause enemy air attackers or observers to search further for the camouflaged real field.

b. Operations Prerequisites. Suitable terrain with adequate drainage must be secured for the decoy airfield.

c. Positioning Decoy in Relation to Landmarks. As far as possible, a decoy airfield should be so situated that its relation to important landmarks and terrain features such as lakes, rivers, railroads, highways, wooded areas, and cultivated fields is similar to that of the real airfield. In some cases it may be practical to simulate some of these necessary landmarks, in dummy fashion, at the same time as the construction of the decoy airfield is progressing.

d. Relation of Decoy to Probable Route of Enemy Approach. As a rule a decoy airfield should be located in the path of the most likely approach an enemy would follow to the real field. The decoy will then be seen by enemy observers before they reach the real field. It is entirely possible, however, that in many cases this consideration in positioning may have to be foregone. The terrain, or existence of prominent landmarks, may indicate a much more desirable site in some other direction from the real field than that of probable enemy approach.

e. Size of Decoy Airfield. A day decoy field should be about the same size as the real field or the standard size of similar known operational fields. This is important because enemy aerial photograph interpreters scale its actual size for comparison with their knowledge of the real field or their knowledge of our normal sized airfields. Substantial divergence in size may cause the field to be suspected and probably identified, upon further reconnaissance, as a fake. Once the decoy is identified as such, air attackers will be briefed on how to avoid it.

f. Aircraft Activity. If the enemy is able to stand off some distance from the suspected airfield and observe it, a complete absence of air traffic would reveal the deception. If intelligence indicates that such a situation is probable, steps should be taken to give a minimum indication of air activity possibly through the use of light planes which could operate from a rough strip.

g. Construction. After the site for the decoy airfield is selected, construction is carried out in the following order:

(1) *Runways.* Since it is the runways, taxiways, and hardstands that make the decoy look like an airfield, they receive first priority in construction. A properly surveyed layout with careful attention to sharp, clear, straight lines of the prominent features is valuable. Usually, the layout can be constructed simultaneously by bulldozing and rough grading. A thin layer of crushed rock, sand, or gravel is sometimes an effective substitute for grading. A well defined ditch will accentuate the outlines of taxiways, runways, and hardstands. For deceiving the air observer, the most important item in decoy airfield construction is the accentuation of these prominent features and any prominent structures, rather than detailed replicas of all features of the real installation. No more detail is necessary than that required to produce the effect of a camouflaged airfield. Grade variations may be permitted; fine grading is never required.

(2) *Access and service roads.* Roads leading to the airfield from main highways, and service roads connecting the hardstands with the bomb storage area, gas storage, and main buildings, have close priority to the actual runways. These roads may be formed in the same manner as the runways and taxiways.

(3) *Gas storage and bomb storage area.* These two facilities are important. The gasoline storage tanks may be simulated by a lumber frame work covered with burlap or by suitable salvage material. Bomb storage pits may be indicated by semicircular parapets of earth pushed up by a bulldozer in a logical position, along the service road.

(4) *Control tower and building.* Control towers and other prominent structures associated with airfields are usually simulated in dummy form in the construction of decoy airfields in rear areas. However, in constructing a decoy of a forward airfield, these structures are seldom erected, since tents or trailers ordinarily serve as the operation centers at forward fields.

(5) *Camouflage.* A decoy airfield should receive the same camouflage as does a real field-simulated terrain patterns on runways and disruptive pattern painting on buildings. Before doing this, it is advisable to study the real airfield by aerial observation and photographs in order to determine the extent to which the decoy airfield should be camouflaged.

68. Supply Points

a. Supply points vary in size and appearance, depending upon their particular function. They do, however, share the common characteristics of extensive tracking, activity, and location adjacent to transportation systems such as harbors, railroads, and road nets. Unless unusual measures are taken, they are usually apparent to even casual aerial observation.

b. A decoy supply point should be near enough to appear to be the real installation and far enough away to allow for possible errors in marksmanship of any attacker. Prominent landmarks must be noted and the decoy located in the same relationship to them that the real installation would be. In some cases the landmarks may be simulated. Figure 78 shows the relationship of several decoy supply points to a real supply point.

c. The decoy must appear to have a convincing road net of the same pattern as the real installation. In addition, troops must be detailed to the decoy site to maintain the appearance of activity. New tracks, scars on the ground, movement of vehicles, and false supplies are essential characteristics for the decoy. If at all possible, it is desirable to route and control through the decoy all traffic to or from the real point. If successful deception is essential, this measure will greatly enhance the decoy's chance of success. In the interest of creating a convincing display it may also be desirable to locate incidental real installations, such as a salvage yard, adjacent to the decoy. In the latter case, the possibility of attack on the decoy should be considered and appropriate protective measures taken. For night deception, certain types of night lighting, such as simulated building lights, showing through tent openings, and decoy fire, are very effective.

69. Oil Refineries and Tank Farms

a. Oil refineries are so difficult to conceal that expenditures for camouflage and decoy construction are not generally justifiable. Night decoys of these installations are, however, quite practicable. In positioning oil refinery decoys, it is very important to follow the same general layout and to place them in surroundings similar to those at the real installations.

b. Tank farms are often included in refineries, but may be separate. If they can be effectively camouflaged, decoys are worthwhile. Construction of a decoy tank farm requires a great deal of effort and ingenuity if the area

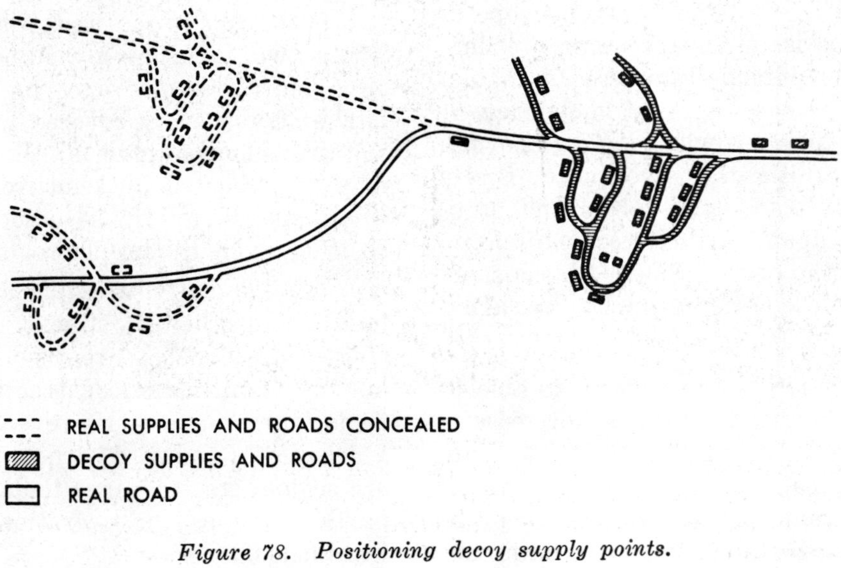

Figure 78. Positioning decoy supply points.

is subject to periodic observation by the enemy. Observation of a decoy under construction obviously robs the decoy of its value. The camouflage of the real farm and erection of the decoy may proceed simultaneously only if enemy observation is improbable. If subject to periodic observation, the decoy should not be revealed until it is completed, interim camouflage being required, and the real farm should not be concealed before the decoy is revealed. Reduction of visibility of the real installation by dark paint and some camouflage may of course precede any activity. Figure 79 depicts a decoy tank farm (A) on one side of a river and two decoy landmarks across the river which correspond to those at the real installation about 3 kilometers away.

70. Railheads

Sidings for unloading supplies, a road net, and storage space are essential facilities for a railhead. Where possible, railheads are established in areas affording the best cover and conceal-

Figure 79. Decoy tank farm and landmarks. Notice that many of the decoy tanks are camouflaged with painted roofs and simulated roads to create the illusion of a housing development. The decoy is camouflaged in the same manner as the real tank farm, but is more obvious. Formerly, a searchlight was near the church on the opposite side of the river from the real installation. At the decoy site a simulated church (B) has been built and the searchlight (C) has been moved from the real site to the decoy site.

ment and may be supplemented by camouflage, air defense facilities and other defensive measures. Supplies are unloaded from railway cars onto trucks and carried to a storage point. Since this is standard practice and known to the enemy, a decoy breakdown point is essential in diverting enemy attacks. It should be more exposed and obvious than the real one.

71. Towns

In flat, barren country or in country with widely dispersed small settlements, it may be desirable, at times, to construct a decoy of the more obvious features of an entire settlement. This might be necessary in the construction of a decoy of an important installation located near such a settlement. Figure 80 shows a decoy town.

72. Night Decoy Installations

a. Introduction. All night decoys employ the same basic equipment; lights, fires, and pyrotechnics. All have the common purpose of confusing the enemy and diverting him from vital targets to areas of little or no importance. The effectiveness of night decoys is dependent upon the following factors: positioning, weather and visibility, proper construction, proper operation and control, maintenance, and camouflage. A decoy may consist of a single or multiple installation, varying in size from a small simulated street light to a complicated installation composed of hundreds of firemaking and lighting devices which, when ignited, will simulate a burning supply depot, factory, airfield, town, or city. There are three types of night decoy installations.

(1) *Type I civil decoys* are those representative of breaches in blackout discipline in factories, warehouses, dock areas, freight yards, towns, and cities.

(2) *Type II airfield decoys* are those representative of runways, marker lights, obstruction lights, wind indicators, and aircraft on the ground.

(3) *Type III field force decoys* are those representative of blackout breaches of supply points, convoys, quarters, and shelters.

b. Factors of Effectiveness.

(1) *Air defense.* Aircraft flying over strange territory at night are strongly attracted by any light, and personnel are often willing to accept any reasonable resemblance of the target as the genuine one. Effectiveness of decoys is proportional to the pressure under which the attack is made. If the air defense is unable to simulate speedy delivery, the chance of successful deception is greatly diminished.

(2) *Enemy pilot.* The determining factor in the design of a night decoy is its appearance to the enemy pilot. He must be able to reasonably identify the target either by vision or by radar and he must be persuaded that it is the target he is seeking.

(3) *Parent target camouflage.* A decoy of a real target will be successful only if the real target is successfully hidden. Since night decoys use light to attract the enemy they are successful only when the surrounding territory is completely blacked out. All measures which make the real target more difficult to locate than the decoy will increase the effectiveness of the deception. Fires started by bombing of the real target must be extinguished as soon as possible. If a fire exists at the target as the result of the attack, further deception operations are not worthwhile against successive waves of attackers unless an equal or more convincing fire at the decoy is started which will have an equal or better chance of enticing followup attackers.

c. Site Selection. The success of a decoy installation will be largely influenced by the adequacy of the site. In the selection of a site the following features should be considered:

(1) The site must be a plausible one. The installation must be in an area where one would expect the target to be located.

(2) The site should provide close similar reference points, discernible at night, to those surrounding the target. To the aerial observer, heavy forests, bodies of water, and open country are distinguishable from each other. Variations in grade up to 8 percent are not distinguishable.

(3) The site should permit the decoy to be oriented in the same compass bearing as the target.

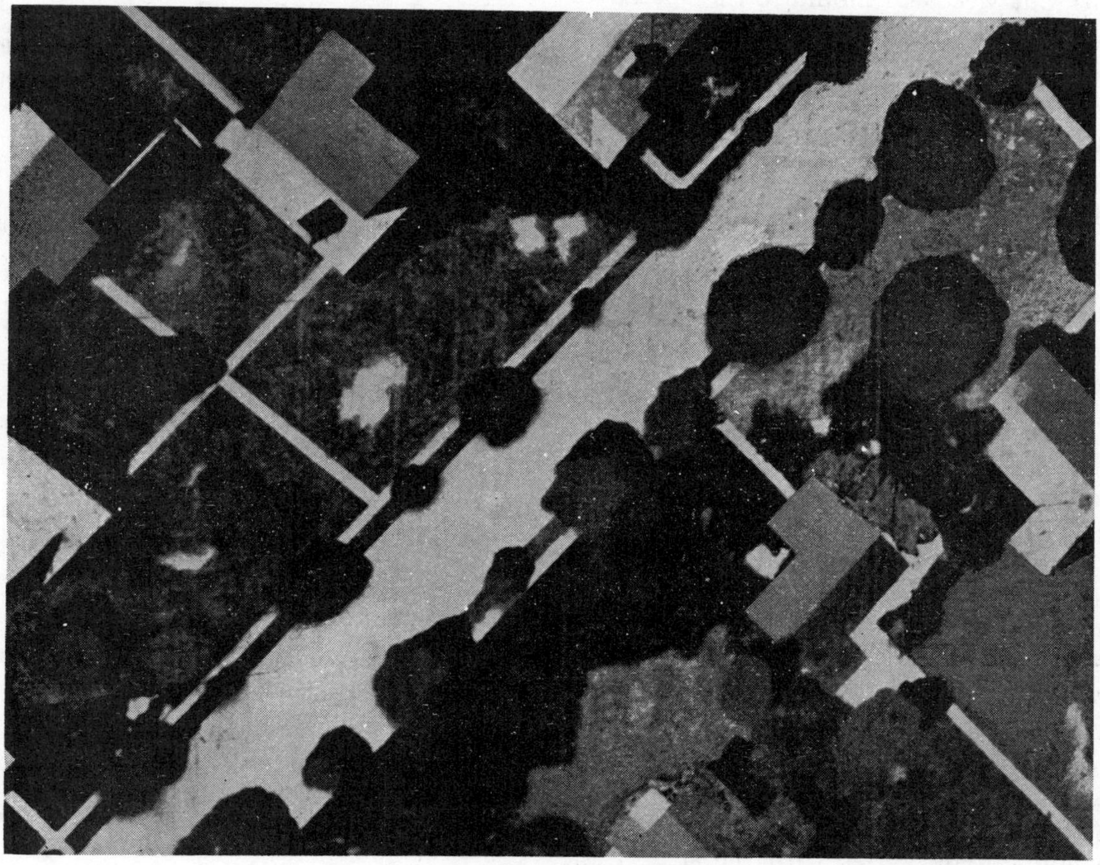

Figure 80. Decoy town.

(4) Where possible, it is desirable to choose a site located along the probable line of approach.

(5) The site must provide an area sufficient for the light pattern on an adequate scale. A three-quarter scale will suffice if the full scale cannot be used.

(6) The site should be within a reasonable distance of the target—close enough to be confused with it, but not so close that the area of poor bombing accuracy overlaps the target or other vital areas.

(7) Wherever possible the decoy should be accessible to adequate roads and power lines.

(8) If a night decoy only, the area should provide easy concealment during daylight hours.

d. Site Preparation. In preparing the site for night decoy installation, it is essential that indications of construction and changes in the appearance of the terrain be kept to a minimum. Normal activities such as farming and grazing should be continued where possible. These precautions not only serve to prevent the enemy from locating and mapping night decoy sites during the day, but also help to preserve local secrecy. The construction will be facilitated if the following procedure is followed:

(1) Establish location of all devices, paying particular attention to service roads and provisions for servicing the equipment.

(2) Remove trees or brush which may constitute a fire hazard, taking care not to scar the landscape.

(3) Cut weeds to minimize the spread of the fire, establishing an 8-foot band of bare earth, 20 to 30 feet in diameter, around each firemaking unit.

e. Materials. The type and amount of equipment used in night decoys will depend upon the nature and scale of the target to be simulated, materials available, and the designers ingenuity. In all instances the cheapest, least critical, and most readily obtainable materials that will serve the purpose should be used.

f. Operation.

(1) In the operation of night decoys, provision must be made for territorial central control and for local control. The central control, which receives immediate information from all sources during an attack on the area, determines which decoy site should be fired and gives the order to the local controller concerned, who passes it on to the detachments in charge of the sites. In the event of a breakdown in communications, the local controller must have authority to act on his own initiative.

(2) The number of men required to operate a night decoy varies with the size of the decoy. The smaller sites require 3 men and the larger displays about 20 men.

(3) Successful operation of a night decoy will require on the part of the operator an understanding of the problem and considerable ingenuity. Variety and "life" are essential. These are obtained by varying the wattages of the lamps used, particularly in "bad blackout" effects, so that as a hostile aircraft approaches closer to a site, the lower powered lighting comes within his range of vision. Furthermore, screens may be erected to obscure observation from certain directions, and skylights may be tilted at various angles so that as an aircraft circles a site the pattern of the decoy is always changing in a life-like manner. Time switches may also be used to switch automatically on and off certain devices such as the open door light, thus introducing "life" without a prohibitive outlay in cable for separate circuits.

(4) Standardization layouts for construction of decoys are not permissible because no two decoys should contain any familiar pattern or regularity. Rather, the designer will have to employ whatever construction the immediate problem requires. Night decoys should be checked and rechecked from the air by an experienced observer. Visibility conditions identical to those under which the enemy will observe the decoy should be chosen. The layout, background, dummy construction, and adjustments of light intensities should be checked under a number of different atmospheric conditions and at different altitudes.

g. Rules to Emphasize.

(1) Do not turn on the decoy lights while under enemy observation.

(2) Be sure to switch off primary lights when the enemy approaches.

(3) Do not switch lights on and off to attract attention.

(4) If enemy fails to bomb, do not switch primary lights back on until the enemy is well out of range.

(5) Do not light large fires until sure that the attack is a raid in force; do not set off decoy for a nuisance raider.

(6) Do not light fires later than 1 hour before dawn.

(7) Remove firing plugs during the daytime to prevent lightning from starting the fires.

(8) If electricity fails use the stand-by generator.

h. Maintenance. On a night decoy maintenance will consist of the following:

(1) Rebuilding and refueling fire devices after a raid. Sufficient material should always be on hand to operate the decoy for three consecutive nights.

(2) Repairing damage after raids and eradicating bomb blast marks.

(3) Cleaning and checking generator.

(4) Checking all wire and contacts periodically.

(5) Lifting and relaying wires if required by the use of the area.

i. Camouflage. In desert or arctic terrain, or in open country where night decoys installation may be discernible during the day, the camouflage of these decoys becomes important unless the decoy is also a day decoy. In any event, the decoy lighting and fire devices will have to be concealed. Simple devices may be covered with a flattop or drape net. The more elaborate devices may be covered with dummy buildings or the installation may be made into a day and night decoy, in which case certain lighting equipment may remain exposed. Some simulator device night decoys in open terrain also have the advantage of greater nighttime deception, because decoy fire in a flat area may produce enough light to reveal the true nature of the site. If the decoy is not a dual purpose day and night decoy, it will be necessary to refill and conceal bomb craters or damage in the area of the night decoy. Indication of inexplicable attacks upon innocent country, evident during day reconnaissance, will immediately identify the site as a night decoy to the enemy observer.

73. Smoke Operations

Smoke has three functions in deception operations:

a. Smoke must be used in conjunction with decoys simulating those installations or situations that would normally produce smoke, such as factories, power plants, and decoy damage.

b. Light smoke must be used in conjunction with night decoys when the visibility and light conditions are such as to expose the decoy. In this situation, white smoke may be used to simulate ground haze or mist.

c. Smoke may be used to screen the site of any activity. Smoke may also be used to simulate activity without the aid of simulated construction. The nature of the decoy-screened activity may be disclosed, apparently unintentionally, by relating it to some other activity or display. This method of effective deception is well adapted to river crossing preparation, beach assault concentration, and assembly points. The ease, economy, and speed with which this type of visual decoy can be implemented are unequalled by other visual methods. Furthermore, if it is properly used and its radar characteristics are reproduced, this type of decoy is almost impossible to identify. The use of smoke has one serious limitation. Strong or rapidly changing winds make the use of smoke difficult, in terms of the operating personnel and the amount of generating equipment required. For details of smoke screening operations, see FM 3–50.

74. Decoy Damage

a. Simulated damage is an especially valuable and practical means of deception for installations which are impractical to conceal. Simulated damage that appears real may induce the enemy to stop or lessen the number and force of his attacks on what he is led to believe is a crippled installation. Decoy damage may be used effectively on oil refineries, railroad sidings, hangars, power plants, bridges, wharves, warehouses, water towers, and other large installations.

b. Damage from bombs and fires is the usual type simulated. Simulated damage is prepared in advance; salvaged material and debris are neatly stacked to conform with existing patterns and are scattered immediately after an enemy attack to simulate bomb hits on the structures. Shallow holes may be dug or blasted to simulate bomb craters and sprayed with waste oil or black paint to appear deep; these are covered until the attack is in progress or until after the attack. During the attack, prepared charges and smoke pyrotechnics may be used and fires ignited. After an attack, the prepared damage is revealed. If deception of this kind is to be effective, speed is essential. Personnel should be trained and organized to follow a well rehearsed drill in the event that the nature of the surrounding area is such that actual new bomb craters away from the installation may compromise the deception. Some provision may be required to conceal these real craters (figs. 81 and 82).

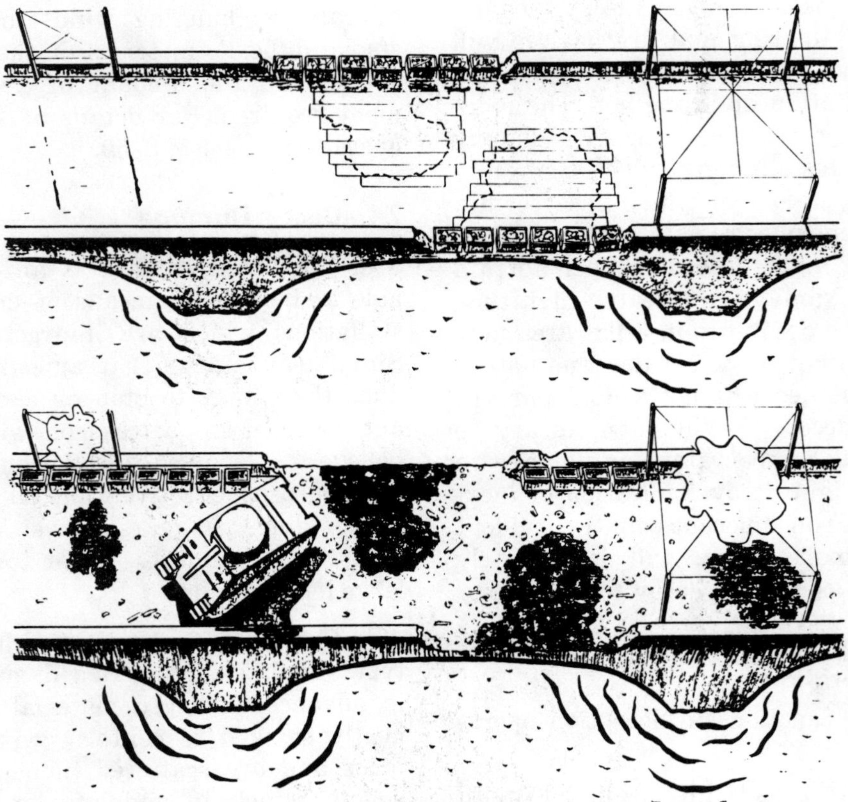

Figure 81. Decoy damage on a bridge.

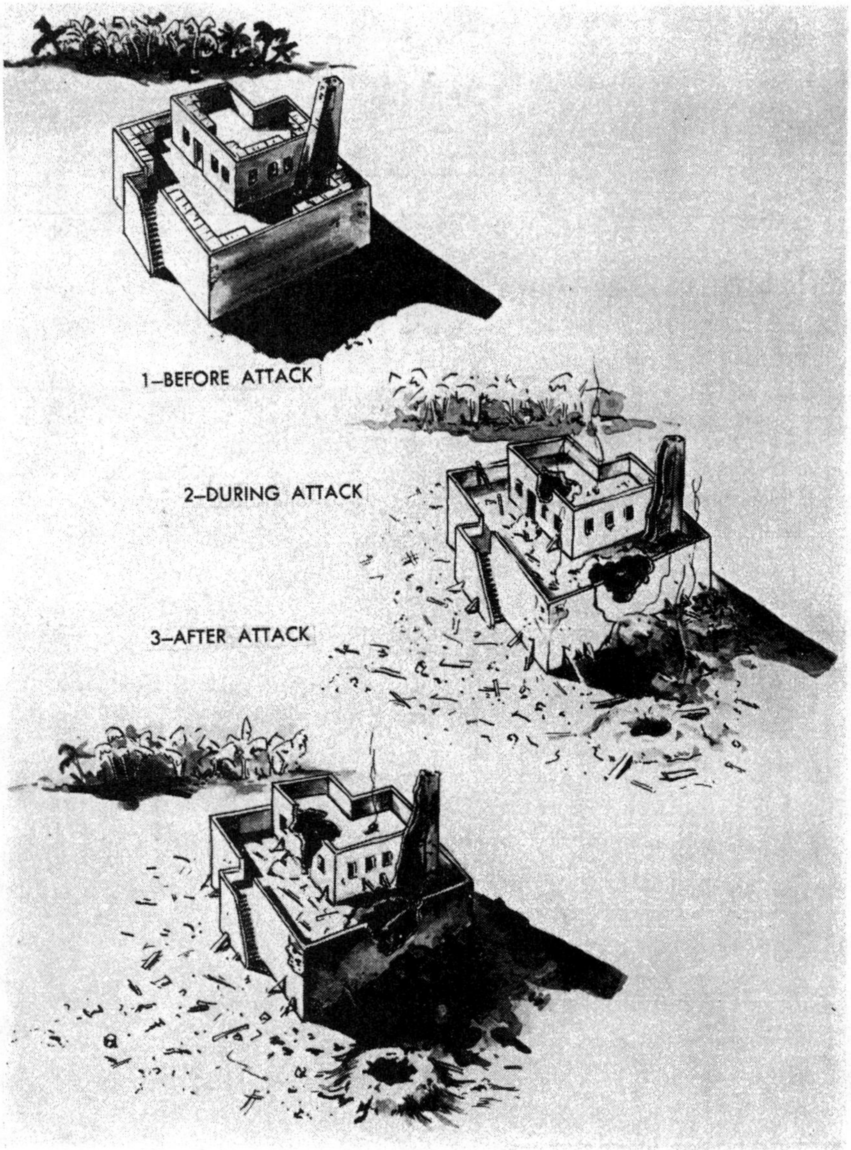

Figure 82. Steps in simulating damage to a building.

APPENDIX

REFERENCES

1. Department of the Army Pamphlets

DA Pam 108–1	Index of Army Motion Pictures, Film Strips, Slides, Tapes, and Phono-Recordings.
DA Pam 310-series	Indexes Pertaining to Administration Training, Maintenance, and Supply.

2. Department of the Army Regulations

AR 320–5	Dictionary of United States Army Terms.
AR 320–50	Authorized Abbreviations.

3. Field Manuals

FM 3–5	Chemical, Biological, and Radiological (CBR) Operations.
FM 5–1	Engineer Operations and Organizations.
FM 5–34	Engineer Field Data.
FM 5–35	Engineer Reference and Logistical Data.
FM 21–5	Military Training.
FM 21–75	Patrolling.
FM 30–10	Terrain Intelligence.

4. Technical Manual

TM 5–200	Camouflage Materials.

INDEX

	Paragraph	Page
Active infrared detecting devices	7*c*	7
Aerial photography	7*a*, 9*b*(2)	4, 15
Aircraft	Ch 7	49
Airfields:		
Concealment	55	84
Simulation	67	92
Artillery:		
Concealment	25–26	44
Simulation	61	90
Audio camouflage discipline	9*b*(3)	17
Barren terrain:		
Bivouac concealment	34	58
Patterns	9*a*(3) (*d*)	15
Bivouacs:		
Concealment	32–35	57
Simulation	62	90
Black and white films	7*a*(4) (*a*)	5
Blackout camouflage discipline	9*b*(2)	15
Blending	10*b*	20
Bridges	52	83
Buildings	49	69
Camouflage detection film	7*a*(4) (*d*)	6
Camouflage discipline	9*b*	15
Canvas equipment	13	26
Clothing	17	28
Color, a factor of recognition	8*e*	9
Color film	7*a*(4) (*b*)	5
Command posts:		
Concealment	36	60
Simulation	63	92
Command responsibilities for camouflage	3*b*	3
Concealment, definition	4	3
Construction, principle of concealment	9*c*	77
Construction of field fortifications	19	32
Covers for foxholes	20	32
Damage simulation	74	99
Deception	10*c*, Ch 10	20, 86
Decoy installations	Ch 10	86
Detection film	7*a*(4) (*d*)	6
Digging in:		
Building construction	49*b*	70
Vehicle concealment	24*e*	38
Direct observation	6	4
Emplacements:		
Camouflage	Ch 5	31
Simulation	60*a*	88

	Paragraph	Page
Factors of recognition	8	7
Far infrared detecting devices	7*c*	7
Field fortifications	Ch 5	31
Films	7*a*(4)	5
Filters for black and white film	7*a*(4) (*a*)	5
Fixed installations	Ch 9	66
Foot tracks simulation	58b(1)	86
Foxhole covers	20	32
Hammocks for aircraft concealment	12	26
Helmet camouflage	12	26
Hiding	10*a*	17
High oblique photographs	7*a*(2)	4
Indirect observation	7	4
Individual camouflage	Ch 4	26
Individual camouflage face paint	22	33
Infrared detecting devices	7*c*	7
Infrared film	7*a*(4) (*c*)	5
Infrared reflectant paint	7*a*(4) (*c*)	5
Installations, large scale	Ch 9	66
Intrenchment simulation	60*a*	88
Layout grid control	45	67
Low oblique photographs	7*a*(3)	5
Machinegun emplacement	21	32
Methods of camouflage	10	17
Minefield simulation	58*h*	88
Modular system	45	67
Mortars	22	33
Motor parks simulation	65	92
Movement, a factor of recognition	8*f*	9
Natural materials	24*b*	36
Near infrared detecting devices	7*c*	7
Nets:		
Aircraft concealment	30*a*	50
Artillery concealment	26*c*	44
Vehicle concealment	24*d*, *e*	38
Night camouflage discipline	9*b*(2)	15
Night decoy installations	72	96
Oblique photographs	7*a*(2), (3)	4, 5
Observation:		
Direct	6	4
Indirect	7	4
Obstacle simulation	60*c*	88
Oil refineries simulation	69	94
Panchromatic film	7*a*(4) (*a*)	5
Packing areas	50	82
Passive infrared detecting devices	7*c*	7

	Paragraph	Page
Pattern painting:		
Aircraft	31	55
Artillery	26*d*	44
Helmets	12	26
Skin	14	27
Vehicles	24*c*	36
Weapons	15	28
Permanent installations	Ch 9	66
Photography	7*a*	4
Pillboxes	60*b*	88
Pipelines	54	84
Position, a factor of recognition	8*a*	8
Principles of concealment	9	14
Quartering party	23*a*(2)	35
Radar detecting devices	7*b*	6
Railheads:		
Concealment	51	82
Simulation	70	95
Refineries	69	94
Revetments	55*ff*	84
Recognition factors	8	7
Relative position, a factor of recognition	8*a*	8
Responsibilities for camouflage	2	3
Roads	50	82
Rolling stock	57	86
Rural terrain patterns	9*a*(3)(*a*)	15
Sandbags	24*e*	38
Screening	10*a*	17
Shadow:		
A factor of recognition	8*c*	8
Aircraft concealment	30*a*	50
Building concealment	49*a*	69
Vehicle concealment	23*c*	36
Shadow nets	30*a*	50
Shape, a factor of recognition	8*b*	8
Shine:	8*d*, 16	8, 28
Aircraft concealment	29	49
Vehicle concealment	23*b*	36
Signatures	58	86
Siting:		
Aircraft	28	49
Artillery	26*b*	44
Field fortification	18	31
Principles of	9*a*	14
Vehicle	24*a*	36
Skin	14	27
Smoke	73	99
Snow camouflage clothing	17	28
Snow covered terrain:	17	28
Bivouac concealment	35	58
Use of nets	24*d*	38
Vehicle concealment	23*a*(4), 23*c*	36
SOP for camouflage	9*b*(1)	15
Sound camouflage discipline	9*b*(3)	17
Sound screens	9*b*(3)	17
Stages of development of a bivouac	33	57
Stereophotography	7*a*(1)	4
Supply points:		
Concealment	37	62
Simulation	68	94
Tank farm simulation	69	94
Tank simulation	66	92
Terrain patterns	9*a*(3)	14
Texture, a factor of recognition	8*d*	8
Town simulation	71	96
Track characteristics	Fig. 72	87
Track discipline	9*b*(1), 23*a*	15, 35
Tracked vehicle track simulation	58*b*(3)	86
Training	23*a*(3)	35
Trains	51	82
Trench simulation	60*a*	88
Troop concentration simulation	64	92
Umbrella screens	30*c*	54
Urban terrain patterns	9*a*(3)(*b*)	15
Vehicles	23–24	35
Vertical photographs	7*a*(1)	4
Water points	38	65
Weapons	15	28
Wheeled vehicle track simulation	58*b*(2)	86
Wire line	52	83
Wooded terrain patterns	9*a*(3)(*c*)	15

By Order of the Secretary of the Army:

HAROLD K. JOHNSON,
General, United States Army,
Chief of Staff.

Official:

KENNETH G. WICKHAM,
Major General, United States Army,
The Adjutant General.

Distribution:

To be distributed in accordance with DA Form 12–11 requirements for Camouflage, Basic Principles and Field Camouflage.